CLARENDON LECTURES IN G
ENVIRONMENTAL STUDIES

General Editors

Gordon Clark   Colin Clarke   Andrew Goudie

The Geographical Structure of Epidemics

PETER HAGGETT

CLARENDON PRESS · OXFORD

OXFORD

UNIVERSITY PRESS

Great Clarendon Street, Oxford OX2 6DP

Oxford University Press is a department of the University of Oxford.
It furthers the University's objective of excellence in research, scholarship,
and education by publishing worldwide in

Oxford New York

Athens Auckland Bangkok Bogotá Buenos Aires Calcutta
Cape Town Chennai Dar es Salaam Delhi Florence Hong Kong Istanbul
Karachi Kuala Lumpur Madrid Melbourne Mexico City Mumbai
Nairobi Paris São Paulo Shanghai Singapore Taipei Tokyo Toronto Warsaw

and associated companies in Berlin Ibadan

Oxford is a registered trade mark of Oxford University Press
in the UK and in certain other countries

Published in the United States
by Oxford University Press Inc., New York

British Library Cataloguing in Publication Data

Data available

Library of Congress Cataloging in Publication Data

Haggett, Peter.
The geographical structure of epidemics / Peter Haggett.
p. cm.—(Clarendon lectures in geographical and environmental studies)
Includes bibliographical references and index.
1. Medical geography. 2. Epidemics. 3. Epidemiology.
4. Social medicine. I. Title. II. Series.
RA792.5 .H34 2000 614.4′2—dc21 00–031359
ISBN 0–19–823363–9
ISBN 0–19–924145–7 (Pbk)

1 3 5 7 9 10 8 6 4 2

Typeset by Hope Services (Abingdon) Ltd.
Printed in Great Britain
on acid-free paper by
Biddles Ltd
Guildford and King's Lynn

For
Torsten Hägerstrand,
geographer and friend

PREFACE

To be invited to give a series of four lectures on 'looking back on your own research field' is to submit any academic to an irresistible temptation to rewrite history. If that invitation coincides with the year in which you are retiring then that provides a still further temptation: namely, to interpret one's own research history in a generous light to show some clear (even logical) path in which one solved problem led unerringly to the next.

I confess that my own story is a very different one. As I show in the first lecture, my work on the geography of epidemics began with an accident and over nearly three more decades has stumbled forwards as a series of generally happy and occasionally productive forays which have taken me far away from the research problems which dominated the first phase of my research career.

One continuous thread in the story is the University of Bristol which has been my West Country base for the last thirty-three years. Moving from there to Cambridge brought an unexpected bonus in that Andrew Cliff who was to have come from Northwestern University to study with me at Cambridge was also diverted to come to Bristol and became my first doctoral student there in 1966. As with all outstanding students, the research supervisor gains far more than he gives. A generation later there are a dozen jointly written books on the shelf and several scores of research papers deriving from a working partnership that brought research results and close family friendship in equal measure. If these essays have a hidden author behind them then it is Andrew's view of epidemics as spatial processes which permeates these pages and his contributions which are paramount.

We both have been equally blessed in those other colleagues with whom we worked on various aspects of the spatial structure theme. Keith Ord, Richard Davies, and Keith Bassett in the early years; Mattthew Smallman-Raynor in more recent years. We were fortunate too in help from the Nuffield Trust and, more recently the Wellcome Trust, in underwriting the modest cost of those studies. Equally, the Australian National University awarded me a visiting fellowship which permitted me to work on epidemics in Fiji. At Geneva the World Health Organization and at Atlanta the

US Centers for Disease Control provided crucial research support. I am grateful to Gerard Ward (at Canberra), Yuri Medvedkov, Norman Bailey, Alan Lopez and John Clements (at Geneva), and Steve Thacker and Donna Stroup (at Atlanta): all have helped with advice and resources at various times over the last thirty years.

Finally, my debt to Oxford. These are the first Clarendon Lectures in Geography and Environmental Studies. First, I must thank the School of Geography at Oxford. It was through the energy and initiative of senior Oxford geographers, notably Gordon Clark, Colin Clarke, and Andrew Goudie that this series was planned and the help of the Press secured. What processes led to my own name coming on the list I shall never know but I hope a longstanding (if unrequited) love affair with Oxford played some small part. Second, my debt to the Oxford University Press for underwriting the cost of presentation and publication. I am grateful to Andrew Schuller and Dominic Byatt who in different roles contributed to the success of the lectures. Third, my debt is to St. Peter's College, Oxford, which welcomed me into its fellowship during my stay in Oxford.

Inevitably, there is a difference between the spoken and written word. This comes through most clearly in terms of illustrations. The original lectures were accompanied by 320 slides, most of them in colour. For this written version I have chosen twelve of these to illustrate each chapter and, inevitably these are in black and white. They represent a selection of maps and diagrams that have been drawn or redrawn over many years by a succession of cartographers at Bristol and Cambridge. I am grateful to Roy Versey and Simon Godden in the earlier years, and particularly to Tim Cliff who in recent years has carried the major part of the burden and who brings imagination and skill to cartographic interpretation.

The lectures were given at the School of Geography in Mansfield Road, Oxford, in November 1998 and written up in the Isle of Purbeck six months later. I am grateful to Clare and Tim Haggett and to Ann and Arthur Tait for making 'The Old Dairy House' available as a base for the writing. The early summer view from the study over the lovely garden and the Dorset landscape beyond provided such a temptation to join Brenda outside, that it served as a spur to finishing the work (more or less) on time.

P.H.

Arne, Dorset
June 1999

ACKNOWLEDGMENTS

I am indebted to my research collaborators Andrew Cliff, Keith Ord, Matthew Smallman-Raynor, and Roy Versey and to our publishers for permission to redraw or reproduce figures which originally appeared in our earlier publications. *Spatial Diffusion* (Cambridge University Press, 1981), Figures 1.1, 1.2, 1.6, 1.8, 2.4, 2.5, and 2.8; *The Spread of Measles in Fiji and the Pacific* (Australian National University Press, 1985), Figures 2.9, 2.10, 2.11, and 2.12; *Atlas of Disease Distributions* (Blackwell, 1988), Figures 1.9, 1.11, 2.2, 2.3, 2.6, 2.7, 3.1, 3.2, and 4.5; *Atlas of AIDS* (Blackwell, 1992), Figures 3.3 and 3.12; and *Measles: an historical geography* (Blackwell, 1993), Figures 2.1, 3.4, 3.6, 4.1. I am also grateful to the American Geographical Society for permission to reproduce Figure 3.5 from C.O. Sauer's Bowman Lectures; to Professor David Bradley of the London School of Hygiene for Figure 3.9; to the Centre for Disease Control for data used in Figure 1.4; to the late Professor Peter Gould of the Pennsylvania State University for permission to redraw Figures 3.8 and 4.9; to Professor R.R. Tinline of Queen's University, Canada, for Figure 4.10; and to the World Health Organization, Geneva, for Figure 4.12. Full identification of all sources are given in the figure captions.

CONTENTS

List of Figures   xiii
List of Tables   xv

1. EPIDEMICS AS DIFFUSION WAVES   1

   Geography and spatial diffusion   1
   The work of Torsten Hägerstrand   4
   Hägerstrand's Monte Carlo diffusion model   6
   Applications to epidemiological data   9
   The nature of epidemics   10
   The burden of communicable disease   12
   Measles as a tracker epidemic disease   16
   Epidemic disease modelling: an historical note   21
   Simple mass–action models   21
   Kendall and spatial waves   26
   Epidemics as spatial diffusion processes   29

2. EPIDEMICS ON SMALL ISLANDS   31

   Islands as natural epidemic laboratories   31
   Iceland as a laboratory   33
   Iceland's record of epidemic waves   37
   Generalizations and wave sequences   43
   Iceland as a graph   46
   Iceland: predicting epidemic measles waves   47
   Measles in Fiji and the south-west Pacific   52
   The Fijian outbreak of 1875   53
   The demographic impact of the epidemic   59
   Indian migration to Fiji   61

3. GLOBAL ORIGINS AND DISPERSALS   67

   The geographical question of disease origins   69
   The limits of historical evidence   74
   Carl Sauer and geographical speculation   77

Extension of Sauer's methods to disease origins   80
Global change and its disease implications   82
Growth and relocation of the human population   85
Changing global land use   87
Global warming   91
The collapse of geographical space   93
Disease implications of global change   95

4. CONTAINING EPIDEMIC SPREAD   99
Spatial control strategies   99
Local elimination and natural breaks in infection   103
The impact of vaccination on epidemic cycles   104
Defensive isolation against epidemics   110
Offensive containment   111
Ring control strategies   115
Global eradication: the smallpox campaigns   119
Global campaigns for other epidemic diseases   121
Poliomyelitis elimination campaigns   126
Conclusions   129

Notes   131
Index   145

LIST OF FIGURES

1.1. Types of spatial diffusion   3
1.2. Economic impulses as diffusion waves   4
1.3. Real and simulated diffusion patterns   8
1.4. Annual mortality and morbidity from major infectious and parasitic
     diseases in the United States in the mid 1980s   15
1.5. Characteristics of a measles epidemic   18
1.6. Measles cycles and population size   20
1.7. Basic elements in the three most elementary Hamer–Soper models of
     wave generation   22
1.8. Bartlett's findings on city size and epidemic recurrence   23
1.9. Conceptual view of the spread of a communicable disease (measles) in
     communities of different population sizes   25
1.10. Kendall model of the relationship between the shape of an epidemic
     wave and the susceptible population/threshold ratio (S/ρ)   27
1.11. Changes in the shape of epidemic waves with distance from the origin
     of an epidemic   28
1.12. Map sequences as a predictive device   30

2.1. Epidemic frequency and population size relationships for island
     communities   33
2.2. Settlement distribution in Iceland   35
2.3. Changes in the boundaries of recording districts for Icelandic
     medical data over an eighty-year period   36
2.4. Spread of wave III of the measles epidemic in north-west Iceland, 1904   40
2.5. Spread of waves VII and XIII of the measles epidemic in Iceland   42
2.6. Time intervals between measles epidemic waves in Iceland   44
2.7. Average lag structure for measles epidemics in Iceland, 1945–70   45
2.8. Seasonal distribution of measles contacts in Iceland   50–1
2.9. Sequence of measles epidemic waves in Fiji, 1875–1982   52
2.10. Generalized directions of measles spread in the Pacific Basin since
     1800   54–5

2.11. Fiji, 1875: elements in the spread of measles in the first six weeks of the epidemic   57

2.12. The impact of changes in transport on disease introduction   63

3.1. Origin and worldwide spread of cholera   70

3.2. Origin and spread of an influenza pandemic   71

3.3. The global diffusion of HIV-1   72

3.4. Historical records of epidemics   75

3.5. Sauer's suggested origins of agriculture in the New World   79

3.6. Hypothetical reconstruction of original hearth of measles and its spread   83

3.7. Global population growth   85

3.8. Transmission of AIDS through the United States urban hierarchy   88–9

3.9. Increasing travel over four male generations of the same family   93

3.10. Increased spatial mobility of the population of France over a 200-year period, 1800–2000   94

3.11. Relative threats posed by communicable diseases to travellers in tropical areas   95

3.12. History of the 1967 Marburg fever outbreak in Europe   97

4.1. Simplified model of control strategies for an infection process   100

4.2. Schematic diagram of four spatial and non-spatial control strategies to prevent epidemic spread   101

4.3. Predicted effects on epidemics of widespread immunization   105

4.4. Time–age predictions of the number of cases of a virus disease following vaccination   106

4.5. Measles reduction in the United States   108

4.6. Pennsylvania measles outbreak, 1987–8: spread in the Amish community   109

4.7. Rubella (German measles) in New South Wales, Australia   112

4.8. Predicted map of rabies spread   114

4.9. Projective modelling of the AIDS epidemic in the United States   116

4.10. Spatial impact of disease control strategies   118

4.11. Global eradication of smallpox under the WHO Intensified Programme, 1967–77   122

4.12. Progress on the eradication of poliomyelitis   128

LIST OF TABLES

1.1. Examples of some major outbreaks of epidemics in world history   11

1.2. Communicable diseases in relation to overall global health   13

2.1. Multiple epidemic waves of measles for Iceland over a 180-year period   38

2.2. Forecasting models applied to measles epidemics in Iceland   47–8

2.3. Estimated death rates from measles epidemics in isolated populations   60

3.1. Examples of emerging virus diseases   68

3.2. Geographical change and virus emergence   84

4.1. Target diseases in the WHO expanded programme on immunization   123

4.2. Biological and sociopolitical features that favoured the global eradication of smallpox   124

4.3. Smallpox, measles, and poliomyelitis: comparison of biological and sociopolitical features favouring global eradication   125

1

Epidemics as Diffusion Waves

In this opening lecture, I wish both to pose and to try to answer three questions. First, what do geographers mean by diffusion waves and why is it a deserving topic for research? Second, what are epidemics and how can an understanding of diffusion concepts help to understand them? Third, how can we build usable models of the way in which epidemics move through geographical space? These basic questions form the foundations from which we go on to explore specific examples of geographical work on epidemics (Lecture 2) and to enquire how geographical work can throw light on the emergence and spread of new diseases (Lecture 3). Finally, in the fourth lecture, we turn to the practical questions of whether geographical concepts have any part to play in the containment of epidemic diseases.[1]

Geography and spatial diffusion

Since these lectures are the first in 'Clarendon Lectures in Geography and Environmental Studies', it is relevant at the outset to state what I take geography to be.[2] For me, geography is most succinctly described as 'the study of Earth's surface as the space within which the human population lives'. By 'Earth's surface' is meant the thin and vulnerable shell, termed the biosphere, only about one-thousandth of the planet's circumference thick, which forms the habitat or environment within which the human population is able to survive. In practice we are talking about an arena which has an area of half a billion square kilometres (70 per cent of it water-covered) and with a thickness of only up to around ten kilometres.

Geography as a discipline is made up of three strands. The first is an emphasis on location. Geographers aim to establish locations accurately through various forms of mapping, and then to analyse the various spatial patterns that can be discerned, often through quantitative analysis. A second strand in the character of geography is an emphasis on ecological relations. The phenomena studied are related to the environments within which they are located. It is worth noting that these human–environment relations may be two-way (e.g. the impact of people on land, as well as of land on people) and may be studied at various geographical scales from that of a small locality up to the globe itself. A third strand is an emphasis on the unique character of particular places on the Earth's surface. This regional dimension in geography fuses the spatial and ecological strands on to a specific area or region.

Within the range of geographical concepts, that of spatial diffusion occupies a central place. The *Oxford English Dictionary* defines *diffuse* as 'to disperse or be dispersed from a centre; to spread widely, disseminate'. However, in the geographical literature, the term *diffusion*, has two distinct usages as shown in Figure 1.1. *Expansion diffusion* is the process whereby a phenomenon of interest (this may be information, a material artefact, a disease), spreads from one place to another. In this expansion process, the item being diffused remains, and often intensifies, in the originating region, but new areas are also occupied by the item in subsequent time periods. *Relocation diffusion* is also a spatial spread process, but the items being diffused leave the areas where they originated as they move to new areas. Figure 1.1 shows how the two processes of expansion and relocation differ and how they may be combined, and illustrates them with an example of a cholera pandemic.

Expansion diffusion occurs in two ways. *Contagious* spread depends on direct contact. This process is strongly influenced by distance because nearby individuals or regions have a much higher probability of contact than remote individuals or regions. Therefore, contagious spread tends to occur in a centrifugal manner from the source region outward. Expansion diffusion may, however, occur in a very different fashion, namely by *hierarchical* spread. This describes transmission through an ordered sequence of classes or places, for example from large metropolitan centres to remote villages. Within socially structured populations, innovations may be adopted first on the upper level of the social hierarchy and then trickle down to the lower levels. *Cascade diffusion* is a term reserved for processes that are always assumed to be downwards from larger to smaller centres.

The geographical spread of economic phenomena are an example of research within a diffusion framework. An outstanding early work was that of the German spatial economist, August Lösch on price changes in space.[3]

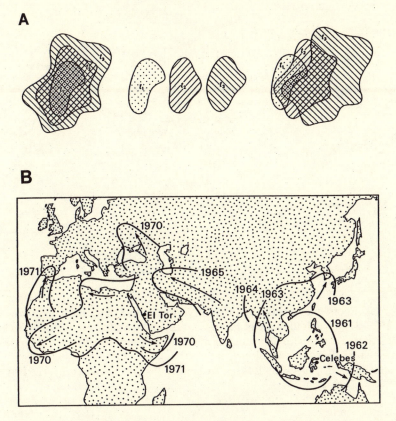

Fig. 1.1. Types of spatial diffusion. (*Above*) Expansion diffusion, relocation diffusion, and combined expansion and relocation processes. The term, t_1, denotes the time period. (*Below*) Example of a combined diffusion wave: the spread of the El Tor cholera pandemic, 1960–71.

Source: Cliff *et al.* 1981, *op. cit.* [note 1], Figs 2.1, 2.2, pp. 6, 7.

Figure 1.2 shows the spread of the business depression of 1929–31 through the state of Iowa, central United States. Graphs show indices of business activity and their running means for ten counties in the state with stipple indicating a period of consistent fall in the running means. The time contours superimposed on the map indicate the time of arrival of the depression front at various places in the state suggesting a ripple-like spread from Chicago, the nearest focus of disturbance, along the main transport arteries, here oriented east–west.

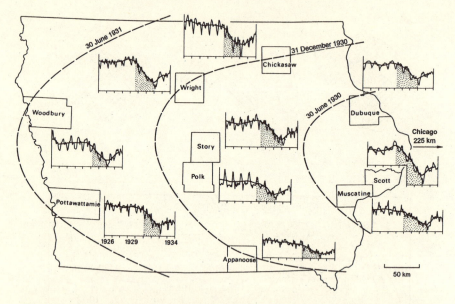

Fig. 1.2. Economic impulses as diffusion waves. Spread of the business depression of 1929–31 through the state of Iowa, central United States. Graphs show indices of business activity with their running means for ten counties. Time contours show time of arrival of the depression 'front'.

Source: Redrawn from Lösch, 1954, *op. cit.* [note 3], Fig. 100, p. 497 in Cliff *et al.* 1981, *op. cit.* [note 1], Fig. 2.6, p. 13.

The work of Torsten Hägerstand

Much of the geographical interest in diffusion studies stems from the work of the Swedish geographer, Torsten Hägerstrand. His *Innovation Diffusion as a Spatial Process*, originally published in 1953, was concerned with the acceptance by farmers of several agricultural innovations, such as the control of bovine tuberculosis by vaccination and subsidies for the improvement of grazing in an area of central Sweden.[4] This book was the precursor of various similar practical studies, particularly in the United States.

In one of his early studies of a contagious diffusion process, Hägerstrand described a four-stage model for the passage of what he termed 'innovation waves' ('*innovations förloppet*'), but which are more generally called 'diffusion waves'. From maps of the diffusion of various innovations in Sweden, ranging from bus routes to agricultural methods, Hägerstrand drew a series of cross-sections to shows the profile of the waveform at different points in time. He suggested that diffusion profiles can be broken into four types, each of which describes a distinct stage in the passage on an innovation through an area.

The *primary stage* marks the beginning of the diffusion process. A centre of adoption is established at the origin. There is a strong contrast in the level of adoption between this centre and remote areas which is reflected in the steep decline of the level of adoption curve beyond the origin. The *diffusion stage* signals the start of the actual spread process; there is a powerful centrifugal effect, resulting in the rapid growth of acceptance in areas distant from the origin and by a reduction in the strong regional contrasts typical of the primary stage. This results in a flattening of the slope of the proportion of adopters curve. In the *condensing stage*, the relative increase in the numbers accepting an item is equal in all locations, regardless of their distance from the original innovation centre; the acceptance curve moves in a parallel fashion. The final *saturation stage* is marked by a slowing and eventual cessation of the diffusion process, which produces a further flattening of the acceptance curve. In this stage, the item being diffused has been adopted throughout the country, so that there is very little regional variation.

Since Hägerstrand's original work, other Swedish geographers have carried out similar studies to test the validity of this four-stage process. For instance, Tornqvist has traced the spread of televisions in Sweden by observing the growth of television ownership in its decade of rapid growth from 1956.[5] Using information obtained from 4,000 Swedish post office districts, he demonstrated that television was introduced into Sweden relatively late, yet within nine years, about 70 per cent of the country's households had bought their first set. Tornqvist's results broadly confirm Hägerstrand's analysis. The diffusion process slowed down, thus indicating the beginning of the saturation phase, at the end of the study period.

The shape of the changing diffusion profile in time and space has been formally modelled. The temporal build-up in the number of adopters of an innovation follows an S-shaped curve when plotted against time, with a logistic curve as the mathematical form most commonly adopted. Casetti and Semple have also considered extensions of the logistic model to handle inhomogeneous mixing as a function of the distance between adopters and potential adopters.[6] Their models may be applied provided that a diffusion pole (i.e. the original centre from which the diffusion process started) can be specified. Cliff and Ord applied a logistic regression model to data on the adoption of tractors in the central farming region of the United States, and have shown how the model may be structured to test hypotheses about the existence of diffusion poles.[7]

The wavefront of adoption formed in space by an innovation may also be described by a logistic curve. Mollison has shown that if the contact probability falls off with increasing distance at least as rapidly as an exponential decay curve, then the diffusion wavefront will migrate steadily across space, and may be approximated by a logistic curve.[8] If, however, the contact probability is not bounded exponentially, the wavefront will disintegrate into a series of spatially erratic contacts.

Hägerstrand's Monte Carlo diffusion model

The first major attempt within geography to formulate a workable model of the *process* of spatial diffusion was that of Hägerstrand. He followed the classic pattern by using a very simple model to establish the nature of the spread process. He then added a sequence of improvements, gradually aproaching reality. In developing his basic model in the context of innovation diffusion, Hägerstrand assumed that the decision of a potential adopter to accept an innovation was based solely on information received orally at face-to-face meetings between the potential adopter and those who had already adopted the innovation. He further assumed that the contact probabilities between tellers and receivers conformed to the neighbourhood effect.[9]

Hägerstrand's model was stochastic in the sense that it was driven by a Monte Carlo random process. It was based on four simple rules. First, the input numbers and spatial locations of adopters in the model were the actual configurations at the start of the diffusion process. For operational convenience the area was divided into a grid of regular cells (e.g. a grid of 5 km × 5 km cells—25 km²). Second, a potential adopter was assumed to accept the innovation as he was told by an adopter. Third, in each iteration of the model every adopter was allowed to contact one other person, adopter or non-adopter. The probability that a carrier would contact an individual located in a particular cell in the study grid was determined by a floating grid, which Hägerstrand called a *mean information field* (MIF). The probabilities for the cells within the MIF were estimated from an analysis of migration and telephone traffic data. The floating grid was placed over each existing adopter in turn, so that the adopter was located in the central cell of the MIF and the contact determined by a stochastic

Monte Carlo process with the probability of contact diminishing with increasing distance. To take into account the reduction in interpersonal communication likely to be caused by communication barriers (e.g. rivers and forests in the Swedish case), a fourth rule was added specifying zero- and half-contact barriers.

Even with this simple model, Hägerstrand was able to perform a series of computer runs to simulate the spatial pattern of acceptance of a subsidy for the improvement of pasture on small farms in the Asby district of central Sweden in the 1920s.[10] An example of this modelling is given in Figure 1.3. But it was clear that modifications to the model to bring it closer to geographical reality would be needed. Three areas of improvement can be noted.

First, the work of Robert Yuill on improving the specification of barrier effects.[11] Yuill envisaged four types of barrier cell in decreasing order of blocking effectiveness: type I, *super absorbing* which absorbed the new contacts and destroyed the transmitters; type II, the *absorbing* barrier, which absorbed new contacts but did not affect the transmitters; type III, the *reflecting* barrier, which absorbed the new contracts, but the transmitters were allowed to make a new contact in the same generation; and type IV, the *direct reflecting* barrier, which did not absorb the contacts but instead reflected them randomly to the nearest available cell. The effect of these modifications was illustrated by the passage of wavefronts through a complex series of barriers. Yuill's work was developed further by Richard Morrill who applied various polynomial, sine, and cosine waves to innovation phenomena, and considered what happens when waves from different centres meet.[12]

A second area of modification relates to the complexity of the space through which a diffusion wave passes. Special attention has been paid to the role of settlement hierachies and the ways in which the initial 'frontier' is led through the urban hierarchy.[13] As Hägerstrand observed, a likely first point of introduction to a new country is its primary city, or sometimes another metropolis. Then, centres next in rank follow. Soon, however, this order is broken up and replaced by one where the neighbourhood effect dominates over the pure size succession.

The third overall change relates to computing power. The narrow range of spatial configurations which could be applied in the 1950s computing context has been replaced by one in which simulation of extremely complex geographical situations is now routine. Results can also be checked and rechecked using billions of calculations.

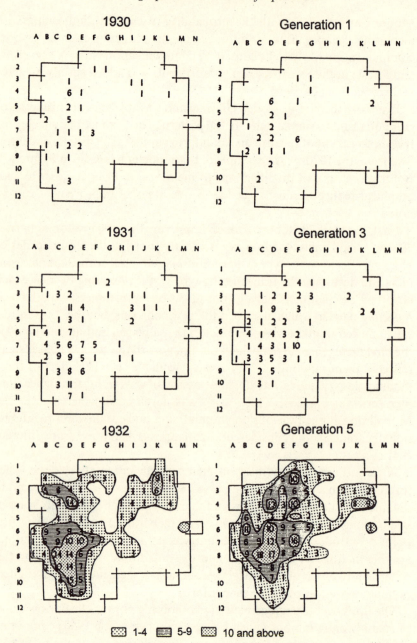

1-4   5-9   10 and above

Applications to epidemiological data

My own interest in Hägerstrand's work dates back to my teaching at Cambridge some forty years ago. After starting at University College London in 1955 I had gone back to Cambridge and developed a course on locational analysis in human geography.[14] One of the elements in the course that gave me considerable stimulation was developing Torsten Hägerstrand's ideas on spatial diffusion—the way in which innovation waves are propagated through human populations.

Hägerstrand had suggested to me that, if I were to work in that area, it might be fruitful to look at the ways in which repetitive waves moved over space (he had worked mostly on single waves) in order to be able to build up, check, and recheck models. My move to Bristol in 1966 with an appointment to the South West Economic Planning Council allowed a rare opportunity to follow up this lead. Specifically, I was trying to extend Lösch's work (see Figure 1.2) on modelling the way in which economic waves, generated by business cycles and measured by local employment data, moved through the network of large cities, small towns, and rural areas in the six counties of south-west England.[15]

Thus, in 1969 I was still working on a conventional research project on economic geography with no thought of epidemics in mind. The telephone call which was to jolt me across on to a quite different track was from an American colleague, Professor Brian Berry in the geography department at the University of Chicago.[16] Brian was calling from home where he was confined to his bed by a severe bout of 'flu. He was due to be in Geneva in two days' time to act as a consultant to a small group of epidemiologists charged with modelling disease spread by the World Health Organization (WHO). The eminent biologist, Marston Bates, had recommended a geographer should be part of a five-man team reviewing the research programme. But Brian was in no condition to travel and the question was:

Fig. 1.3. Real and simulated diffusion patterns. Simulation of the spatial pattern of the number of adopters of improved pasture subsidy in central Sweden. From the initial distribution of adopting farmers in 1929 (generation 0) the next three years are shown; the actual distribution on the left (1930, 1931, 1932) and the simulated distribution on the right (generations 1, 3, and 5). Each square cell on the map is 5 km × 5 km on the ground and numbers give the number of adopters in each cell.

Source: Hägerstrand, 1967, *op. cit.* [note 4], Fig. 3, p. 23. This version is redrawn from Yeates, 1974, *op. cit.* [note 4].

'Would I substitute for him and occupy the geographer's slot on the advisory panel?' I hesitantly agreed, rescheduled my teaching, and set off the next morning for Switzerland. I little knew that I was establishing a chain of serendipitous events[17] which would lead to thirty years of research on epidemics.

However, the visit to Geneva proved to be no immediate Pauline conversion. The WHO team set out their programme and the five-man team each made their comments; me struggling to master the papers and trying hard to conceal my huge ignorance of epidemic processes. The rather brief report with recommendations was agreed and we dispersed to our several countries.[18] But three links made that week were to prove fruitful, and Norman Bailey (WHO), Geoff Watson (Princeton), and Oliver Lancaster (Sydney) all saw a potential for the geographical modelling of disease that I had failed to observe. Exchanges of letters and notes followed, and within two years I had worked at the United States Centers for Disease Control at Atlanta on a WHO Fellowship, and the models which had tracked multiple-wave economic cycles in the south-west were now being adapted to analyse multiple-wave disease cycles. Berry's 'flu virus was having some unexpected sequelae on the other side of the Atlantic.

The nature of epidemics

But what exactly were epidemics? The term *epidemic* comes from two Greek words: *demos* meaning 'people' and *epi* meaning 'upon' or 'close to'. It was used around 500 BC as a title for one major part of the Hippocratic corpus, but the section concerned was mainly a day-to-day account of certain patients and not an application of the word in its modern sense. In addition to its wider usage in terms of public attitudes (e.g. Burke's 'epidemick of despair'), the word has been used in the English language in a medical sense since at least 1603 to mean an unusually high incidence of a disease where 'unusually high' is fixed in time, in space and in the persons afflicted as compared with previous experience. Table 1.1 shows some examples of epidemics in history.

The *Oxford English Dictionary* defines an epidemic as: 'a disease prevalent among a people or community at a special time, and produced by some special causes generally not present in the affected locality'. The parallel term, *epizootic*, is used to specify a disease present under similar conditions in a non-human animal community.

Table 1.1. Examples of some major epidemic outbreaks of disease in world history

Size	Time period	Location	Disease	Estimated number of deaths	Estimated ratio of deaths to population
>1 million deaths	1346–52	Western Europe	Black Death, bubonic plague	20,000,000	1 : 4
	1918–19	Worldwide	Influenza A	20,000,000	1 : 25 (India)
>100,000 deaths	1741	Ireland	Famine, typhus, dysentry	300,000	1 : 6
	1098–9	Palestine (1st Crusade)	Epidemic diseases, famine	240,000	1 : 1.25
	1781–2	Europe	Influenza	100,000	?
>10,000 deaths	c.1438	Paris	Smallpox	50,000	1 : 4
	1870–1	Paris (siege)	Smallpox	75,167	1 : 29
	1870	England & Wales	Scarlet fever	36,000	1 : 650
	1875 (Jan–June)	Fiji	Measles & sequelae	30,000	1 : 4
	1801–3	Haiti	Yellow fever	22,000	1 : 1.13

Source: A. D. Cliff, P. Haggett, and M. Smallman-Raynor, *Deciphering Global Epidemics*. Cambridge: Cambridge University Press, 1999. Table 1.4, pp. 18–21. The original extended table gives, for each epidemic, a detailed list of sources on which the table is based.

In the standard handbook of human communicable diseases, Benenson defines an epidemic more fully as:

The occurrence in a community or region of cases of an illness (or an outbreak) clearly in excess of expectancy. The number of cases indicating presence of an epidemic will vary according to the infectious agent, size and type of population exposed, previous experience or lack of exposure to the diseases, and time and place of occurrence; epidemicity is thus relative to usual frequency of disease in the same area, among the specified population, at the same season of the year.[19]

Benenson's account goes on to stress that what constitutes an epidemic does not necessarily depend on large numbers of cases or deaths. A single case of a communicable disease long absent from a population, or the first invasion by a disease not previously recognized in that area, requires immediate reporting and epidemiological investigation. Two cases of such a disease associated in time and place are taken to be sufficient evidence of transmission for an epidemic to be declared.

Epidemics of communicable disease are of two main types. A *propagated* epidemic is one that results from the chain transmission of some infectious agent. This may be directly from person-to-person as in a measles outbreak, or indirectly via some intermediate vector (malaria) or a microparasite. In some cases, indirect transmission may occur via humans (as in louse-borne epidemic typhus fever, or in a mosquito–man–mosquito chain with malaria). In others, the survival of the parasite is independent of man (thus, *Pasteurella pestis*, the cause of bubonic plague is continually propagated through rodents and the infection of man by an infected flea is in this respect an accidental diversion).

The second type of epidemic is a *common-vehicle epidemic* which results from the dissemination of a causative agent. In this case, the epidemic may result from a group of people being infected from a common medium (typically, water, milk, or food) which has been contaminated by a disease-causing organism. Examples are provided by cholera and typhoid.

The burden of communicable disease

Given the overall decline in mortality from infectious diseases over the course of this century, do epidemic diseases still merit study? The answer is an unequivocal 'yes'. If today we turn the leaves of a single week's report by an epidemiological agency, we find a rich range of disease outbreaks being reported. Thus, Australia's *Communicable Diseases Intelligence* for the week ending 4 April 1994 (a week chosen at random) records over 120 different diseases and agents: these include a major mumps outbreak in western Australia, 131 cases of hepatitis C, an outbreak of Ross River virus infection in the Northern Territory, while its overseas section records Japanese B encephalitis breaking out in Sri Lanka and a severe malaria outbreak on the Trobriand Islands off Papua New Guinea. Meantime influenza A was sweeping through twenty-one Russian cities and cholera continued to invade northern Mozambique.

As Andrew Cliff and I have shown in the *Atlas of Disease Distributions* the geographical distribution of communicable diseases has shown major changes over human history.[20] Some common infectious diseases, such as measles, probably did not emerge until around 5,000 years ago when human populations became large and settled enough to maintain infection chains and when prolonged contact with animal populations with measles-related viruses allowed transfer. Still other diseases and conditions have died out in past centuries and continue to do so today.[21]

Infectious and parasitic diseases occupy the Ninth International Code of Diseases (ICD) codes 1–113, but at least 125 additional specific ICD codes reflect infection. Infectious diseases like measles represent a condition caused by a single virus attack, but some codes cover situations where infections may be secondary to other inciting events (e.g. peritonitis). Despite decades of research and data collecting, establishing the precise impact of infectious diseases at the world level remains elusive. Here, we take our estimates from the World Health Organization's *World Health Report 1995* but recognize that the figures given there can be little more than rough estimates.[22]

There are several ways in which an estimate may be made. One approach to measuring communicable diseases is through the deaths they cause—the mortality approach. Table 1.2 shows that infectious diseases occupy half of the top ten places as global killers.

Taken together, infectious diseases and parasites take 16.4 million lives a year, ahead of heart disease which kills 9.7 million. Table 1.2 also indicates another way to measure disease is through disease incidence—the number of new cases of a disease each year. Again, the table shows the dominant position of the infectious diseases (but note that the figures here are in

Table 1.2. Communicable diseases in relation to overall global health[a]

Rank	Deaths Disease/Condition	Number of deaths (000s)	Incidence Disease/Condition	Number of cases (00,000s)
1	Ischaemic heart disease	4,283	Diarrhoea under age 5, including dysentery	18,210
2	Acute lower respiratory infections under age 5[b]	4,110	Acute lower respiratory infections under age 5	2,483
3	Cerebrovascular disease	3,854	Occupational injuries due to accidents	1,200
4	Diarrhoea under age 5, including dysentry	3,010	Chlamydial infections (sexually transmitted)	970
5	Chronic obstructive pulmonary diseases	2,888	Trichomoniasis	940
6	Tuberculosis	2,709	Gonoccocal infections	780
7	Malaria	2,000	Occupational diseases	690
8	Falls, fires, drowning, etc.	1,810	Measles	452
9	Measles	1,160	Whooping cough	431
10	Other heart disease	1,133	Genital warts	320

[a] Estimates based on 1993 figures.
[b] Estimates for some diseases may contain cases that have also been included elsewhere; e.g. estimates for acute lower respiratory infections and diarrhoea include those associated with measles, pertussis, malaria and HIV.
Source: World Health Organization, 1995, *op. cit.* [note 22], Table 1, p. 3.

100,000s compared to 1,000s for deaths). Diarrhoea in children under 5 years account for 1.8 billion episodes a year (and claim the lives of 3 million children). Acute lower respiratory conditions in children, sexually transmitted diseases, measles, and whooping cough remain major problems.

Yet other ways of measuring the disease burden are in terms of prevalence—the total number of people with a given condition—or the burden of disability that a disease causes. Global figures are hard to find and we know all too little about some major infectious diseases. Such fragments of information as we have at the world scale underscore the role of communicable diseases, for example, schistosomiasis (bilharziasis) has a prevalence of some 200 million worldwide, while 10 million are permanently disabled by paralytic poliomyelitis. Still further calculations can be adopted which allow for modifications in the age of attack, giving a heavier weighting to diseases that strike earlier in life.

An alternative approach to estimating the size of the disease burden is to use figures for a single country. For the United States, the Carter Center estimate for the United States that 83 per cent of all deaths from infections occur outside the 'classic' ICD disease codes 1–113: infections and infection-related deaths are important contributors to circulatory, respiratory, and gastrointestinal disease, to infant mortality and morbidity, and to arthritis.[23] Infections complicate a wide range of injuries and have been found to cause malignancies in humans.

It is thus difficult to obtain a clear statistical picture of the real role of infectious diseases in causing mortality and morbidity. For this reason the Carter Center Health Policy Project reworked the available statistics for the United States and compared this with Centers for Disease Control (CDC) Survey Data to provide a revised estimate of the effects of infectious diseases on morbidity and mortality. They concluded that 740 million symptomatic infections occur annually in the United States, resulting in 200,000 deaths per year. Such infections result in more than $17 billion annually in direct costs, not including costs of deaths, lost wages and productivity, and other indirect costs. About 63,000 deaths are currently prevented annually and a further 80,000 deaths could be prevented by using current or soon to be available interventions.

Figure 1.4 summarizes the Carter Center findings. This plots on a logarithmic scale the number of deaths and number of cases for the main infectious diseases. Note that only diseases causing more than 10 deaths or more than 1,000 cases per year are shown. The largest number of deaths (32,000) are caused by pneumoccocal bacteria followed by nosocomial deaths in

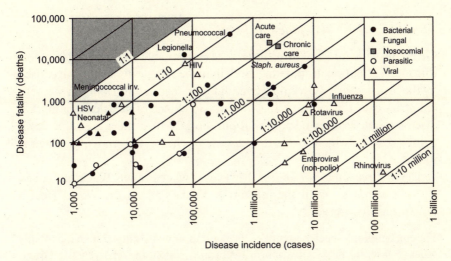

Fig. 1.4. Annual mortality and morbidity from major infectious and parasitic diseases in the United States in the mid 1980s. Note that both disease fatality and disease incidence are plotted on logarithmic scales. The diagonal lines represent the fatality/case ratio.

Source: drawn from data in Bennett, *et al.* 1987, *op. cit.* [note 23], Table 1, pp. 104–7.

acute care (26,400) and chronic care (24,700). In terms of morbidity the largest number of cases are generated by the rhinovirus (125 million) that causes the common cold followed by another group of viruses, influenza (20 million).

The figure shows only the leading 56 of the 117 specific infections used by the Carter Center. Although those not shown have fewer than 10 deaths per year or fewer than 1,000 cases, they include many diseases that rank highly on the 'dread' factor. For example, amoebic meningoencephalitis was recorded only four times in the United States in the year studied but each resulted in death: rabies killed all 10 of those infected, half those 100 cases infected with the cryptosporidiosis parasite died. The case fatality ratio is shown in Figure 1.4 for the more frequently occurring diseases by the diagonal lines which represent fatality–case ratios. Those shown on or above the 1 : 10 diagonal include human immonudeficiency virus (HIV), legionella, meningococcal invasions, and neonatal herpes simplex virus (HSV).

Measles as a tracker epidemic disease

Given the wide range of infectious diseases available for study, it is notable that much attention in epidemic modelling on a single disease is that caused by the measles virus. Given the overall fall in measles mortality in Western countries over this century, the widespread choice of measles as a marker disease might well seem somewhat puzzling. In fact, there are seven compelling reasons why it forms the 'disease of choice' for studying epidemic waves.

The first reason is *virological*. Measles is what Francis Black has called '. . . the simplest of all the infectious diseases'.[24] As the World Health Organization observes:

> The epidemiological behaviour of measles is undoubtedly simpler than that of any other disease. Its almost invariably direct transmission, the relatively fixed duration of infectivity, the lasting immunity which it generally confers, have made it possible to lay the foundations of a statistical theory of epidemics.[25]

It is, therefore, a disease whose spread can be modelled more readily than others.

Although written records of what appear to have been measles outbreaks go back to the Roman period, the causative agent was not identified until the middle of the twentieth century. In 1954, J. F. Enders and T. C. Peebles isolated the cause of measles as a virus; that is, a microorganism that multiplies only within living susceptible cells. Since such viruses range in size down to 20 nm in diameter, an understanding of their basic structure was dependent on the development of the electron microscope. The measles virus was found to be one of a family of viruses (the paramyxovirus family) which produces illnesses in man and other animals. In addition to measles, the family includes canine and phocine distemper, parainfluenza, and rinderpest. As far as present knowledge extends, the measles virus is not thought to undergo significant changes in structure. This assumption is strengthened by the fact that although laboratory research has produced measles viruses with attenuation—decreased virulence—no changes in basic type have yet been recorded.

The way in which measles epidemics occur and propagate in waves, is illustrated in Figure 1.5. This shows that measles has a simple and somewhat regular transmission mechanism that allows the virus to be passed from person to person. No intermediate host or vector is required. The explosive growth in the number of cases that characterizes the upswing of a major epidemic implies that the virus is being passed from one host to many others.

Second, *epidemiological* reasons. Measles exhibits very distinctive wave-like behaviour. Figure 1.6 shows the time series of reported cases between 1945 and 1970 for four countries, arranged in decreasing order of population size. In the United States, with a population of 210 million in 1970, epidemic peaks arrive every year, and in the United Kingdom (56 million) every two years. Denmark (5 million) has a more complex pattern, with a tendency for a three-year cycle in the latter half of the period. Iceland (0.2 million) stands in contrast to the other countries in that only eight waves occurred in the twenty-five-year period, and several years are without cases.

Third, *clinical* reasons. The disease can be readily identified with its distinctive rash and the presence of Koplik spots within the mouth. This means accurate diagnosis without the need for expensive laboratory confirmation. There was early confusion with smallpox, and later with rubella and scarlatina (scarlet fever). Not only does measles display very high attack rates but, crucially, the relative probability of clinical recognition of measles is also high with over 99 per cent of those infected showing clinical features. Thus, in clinical terms, measles is a readily recognizable disease with a low proportion of both misdiagnosed and subclinical cases.

Fourth, *statistical* reasons. The high rate of incidence leads to very large number of cases. Even with under-reporting, major peaks are clearly identified. Measles is highly contagious with very high attack rates in an unvaccinated population. It generates, therefore, a very large number of cases over a short period of time to give a distinct epidemic event. This high attack rate is supported by the many reliable estimates in the literature of the proportion of a population that has had measles. In an early paper, Chapin cites figures for several cities—Aberdeen, Scotland, 90–95 per cent; Willesden, London, 93 per cent; and Providence, Rhode Island, 96 per cent—and concludes that '. . . it is probable that in England, Canada, and the USA [the countries he studied] over 90 percent of urban populations contract measles at some time during their lives'.[26] With vaccination that proportion has fallen. In contrast to the 1920s, analysis of Icelandic populations suggests that because of vaccination the proportion of the population infected with measles by their twentieth birthday can now be as low as 35 per cent. While the precise level will vary both historically and regionally, the very high incidence rates in comparison with other infectious diseases give higher confidence in the reported data than with other low incidence diseases.

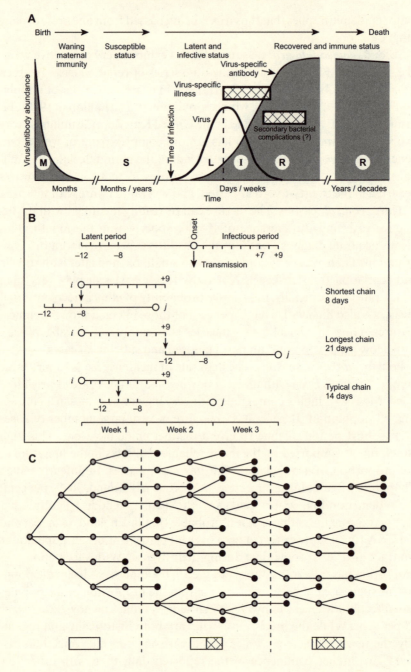

Fifth, *geographical* reasons. The disease is as widespread as the human population itself is in the early twenty-first century. This global potential does not mean that there are not significant spatial variations. Measles in isolated communities, which are rarely infected, has a very different temporal pattern from those in large metropolitan centres where the disease is regularly present.

Sixth, *mathematical* reasons. The regularity has attracted mathematical study since at least 1888 when D'Enko carried out his studies of the daughters of the Russian nobility in a select St. Petersburg boarding school. Hamer has played a major part in testing of mathematical models of disease distribution, most notably in chaos models. It is significant that when Cambridge University set up its prestigious Isaac Newton Centre for the study of mathematics, a session on epidemic models was included in its first year of operation.

Finally, the study of measles is driven by *humanitarian* reasons. Despite major falls in mortality over this century, it still remains a major killer. It accounts for nearly 2 million deaths worldwide, mainly of children in developing countries. As we see in the final lecture, it is on the World Health Organization list for eventual global elimination and substantial progress has been made in the United States with its 'measles zero' campaign. In addition to our basic requirements for a study of disease diffusion, we wished to examine a disease for which the results would have practical ramifications for control. Measles also fits this requirement. Like smallpox, the measles virus is theoretically eradicable. Study of the spatial structure of this particular disease is therefore likely to be of use in planning future eradication campaigns.

Fig. 1.5. Characteristics of a measles epidemic. (A) Disease spread at the individual level. Typical time profile of infection in a host individual. Note the time breaks and different scales for time duration within each phase of the overall lifespan (M, maternal protection; S, susceptible; L, latent; I, infectious; R, recovered). (B) The infection process as a chain structure. The average chain length of 14 days is shown. (C) Burnet's view of a typical epidemic where each circle represents an infection, and the connecting lines indicate transfer from one case to the next. Black circles indicate individuals who fail to infect others. Three periods are shown, the first when practically the whole population is susceptible; the second at the height of the epidemic; and the third at the close, when most individuals are immune. The proportion of susceptible (white) and immune (hatched) individuals are indicated in the rectangles beneath the main diagram.

Source: (A) Haggett, 1994, *op. cit.* [note 1], Fig.1, p. 6. (B) Cliff *et al.* 1993, *op. cit.* [note 1], Fig. 1.3, p. 6. (C) Burnet and White, 1972, *op. cit.* [note 31], Fig. 14, p. 128.

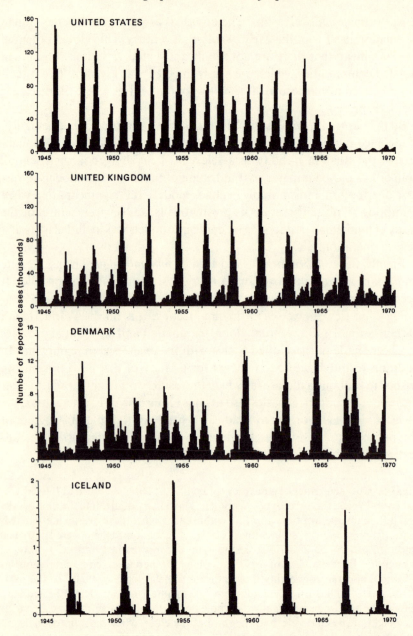

Fig. 1.6. Measles cycles and population size. Reported cases of measles per month, 1945–70, for four countries arranged in descending order of population size. Note the characteristic cyclicity in all cases, the dramatic reduction in amplitude for the United States after 1964 (because of vaccination programmes), and the fact that only Iceland has clear non-endemicity.

Source: Cliff *et al.* 1981, *op. cit.* [note 1], Fig. 3.1, p. 39.

In this section I have tried to show that the choice of measles, within the range of infectious diseases, for geographical study is soundly based. Its analysis gives better prospects for: (a) formulating and testing hypotheses, (b) the developing of models of the geographical diffusion of infectious disease, and (c) using such models in a way that promises to increase our understanding of the spread of a virus whose eradication as a public health problem is being actively sought.

Epidemic disease modelling: an historical note

Interest in the geographical distribution and spread of diseases around the globe is of some antiquity. Hirsch's *Handbuch der historisch-geographische-Pathologie* was first published 150 years ago.[27] However, the evidence of mathematical modelling of epidemic processes goes back still further. Among the first applications of mathematics to the study of infectious disease was that of Daniel Bernoulli in 1760 when he used a mathematical method to evaluate the effectiveness of the techniques of variolation against smallpox.[28] As disease statistics accumulated during the early nineteenth century so William Farr in 1840 was able to fit a normal curve to smoothed quarterly data on deaths from smallpox in England and Wales over the period 1837–9.[29] This empirical curve-fitting approach was further developed by John Brownlee who considered in detail the 'geometry' of epidemic curves.[30]

The origin of modern mathematical epidemiology owes much to the work of En'ko, Hamer, Ross, Soper, Reed, Frost, Kermack, and McKendrick who, using different approaches, began to translate specific theories about the transmission of infectious disease into simple, but precise, mathematical statements and to investigate the properties of the resulting models. A summary of their work is given by Bailey and by Anderson and May in their definitive accounts.[31]

Simple mass–action models

The simplest form of an epidemic model, the Hamer–Soper model, is shown in Figure 1.7.[32] It was originally devloped by Hamer in 1906 to describe the recurring sequences of measles waves affecting large English cities in the late Victorian period and has been greatly modified over the last fifty years to incorporate probabilistic, spatial, and public health features.

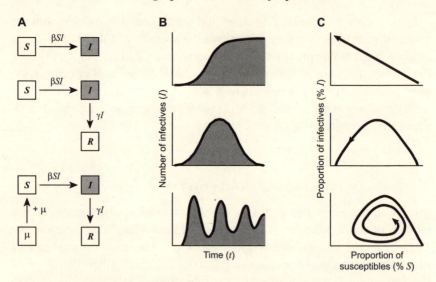

Fig. 1.7. Basic elements in the three most elementary Hamer–Soper models of wave generation. (A) Elements in the model: S, susceptibles; I, infectives; R, recovereds; μ, births; β, infection (diffusion) rate; and γ, recovery rate. (B) Typical time profiles for the model with number of infectives are stippled. (C) Trajectory of typical waves in infective–susceptible space.

The basic wave-generating mechanism is simple. The infected element in a population is augmented by the random mixing of susceptibles with infectives ($S \times I$) at a rate determined by a diffusion coefficient (b) appropriate to the disease. The infected element is depleted by recovery of individuals after a time period at a rate controlled by the recovery coefficient (c). As Figure 1.7 shows, the addition of parameters to the model allows successively more complex models to be generated. A second set of epidemic models based on chain frequencies has been developed in parallel with the mass–action models.

Empirical validation of the mass–action models for measles was provided by the work of statistician Maurice Bartlett. He investigated the relationship between the periodicity of measles epidemics and population size for a series of urban centres on both sides of the Atlantic.[33] His findings for British cities are summarized in Figure 1.8. The largest cities have an endemic pattern with periodic eruptions (Type A), whereas cities below a certain size threshold have an epidemic pattern with fade-outs. Bartlett found the size threshold to be around a quarter of a million:

The critical community size for measles (the size for which measles is as likely as not to fade out after a major epidemic until reintroduced from outside) is found for the United States to be about 250,000 to 300,000 in terms of total population. These figures agree broadly with the English statistics, provided notifications are corrected as far as possible for unreported cases.[34]

Subsequent research has shown that the threshold for measles, or indeed any other infectious disease, is likely to be somewhat variable with the level influenced by population densities and vaccination levels. However, the threshold principle demonstrated by Bartlett remains intact.

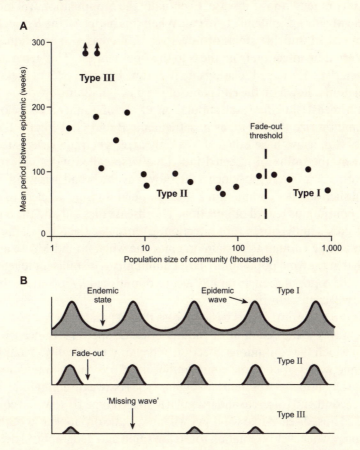

Fig. 1.8. Bartlett's findings on city size and epidemic recurrence. (A) The impact of population size on the spacing of measles epidemics for 19 English towns. (B) Characteristic epidemic profiles for the three types indicated in (A).
Source: Cliff *et al.* 1981, *op. cit.* [note 1], Fig. 3.2, p. 40.

Therefore, once the population size of an area falls below the threshold, when the disease concerned is eventually extinguished, it can only recur by reintroduction from other reservoir areas. Thus, the generalized persistence of disease implies geographical transmission between regions as shown in Figure 1.9.[35] We can see that in large cities above the size threshold, like community A, a continuous trickle of cases is reported. These provide the reservoir of infection which sparks a major epidemic when the susceptible population, S, builds up to a critical level. This build up occurs only as children are born, lose their mother-conferred immunity and escape vaccination or contact with the disease. Eventually the S population will increase sufficiently for an epidemic to occur. When this happens, the S population is diminished and the stock of infectives, I, increases as individuals are transferred by infection from the S to the I population. This generates the characteristic D-shaped relationship over time between sizes of the S and I populations shown on the end plane of the block diagram.

With measles, if the total population of a community falls below the 0.25-million size threshold, as in settlements B and C of Figure 1.9, epidemics can, as we have noted above, only arise when the virus is reintroduced by the influx of infected individuals (so-called index cases) from reservoir areas. These movements are shown by the broad arrows in Figure 1.9. In such smaller communities, the S population is insufficient to maintain a continuous record of infection. The disease dies out and the S population grows in the absence of infection. Eventually, the S population will become large enough to sustain an epidemic when an index case arrives. Given that the total population of the community is insufficient to renew by births the S population as rapidly as it is diminished by infection, the epidemic will eventually die out.

It is the repetition of this basic process that generates the successive epidemic waves witnessed in most communities. Of special significance is the way in which the continuous infection and characteristically regular type I epidemic waves of endemic communities break down, as population size diminishes, into first, discrete but regular type II waves in community B and then, second, into discrete and irregularly spaced type III waves in community C. Thus, disease-free windows will automatically appear in both time and space whenever population totals are small and geographical densities are low.

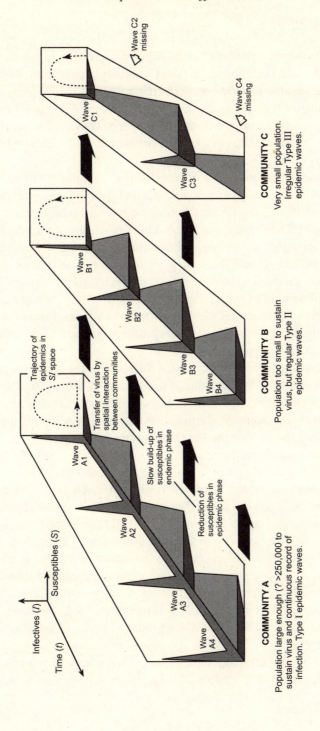

Fig. 1.9. Conceptual view of the spread of a communicable disease (measles) in communities of different population sizes. Stages in spread correspond to the Bartlett model.

Source: Cliff and Haggett, 1988, *op. cit.* [note 20], Fig. 5.5A, p. 246.

Kendall and spatial waves

The relationship between the input and output components in the wave-generating model has been shown by the Cambridge statistician, David Kendall, to be critical.[36] If we measure the magnitude of the input by the diffusion coefficient (b) and the output by the recovery coefficient (c) then the ratio of the two c/b defines the threshold (given by the Greek letter rho, ρ) in terms of population size. For example ,where c is 0.5 and b is 0.0001, then ρ would be estimated as 5,000.

Figure 1.10 shows a sequence of outbreaks in a community where the threshold has a constant value and is shown therefore as a horizontal line. Given a constant birth rate, the susceptible population increases and is shown as a diagonal line rising over time. Three examples of virus introductions are shown. In the first two, the susceptible population is smaller than the threshold ($S > \rho$) and there are a few secondary cases but no general epidemic. In the third example of virus introduction the susceptible population has grown well beyond the threshold ($S > \rho$); the primary case is followed by many secondaries and a substantial outbreak follows. The effect of the outbreak is to reduce the susceptible population as shown by the offset curve in the diagram.

Kendall has investigated the effect of this S/ρ ratio on the incidence and nature of epidemic waves. With a ratio of less than one, a major outbreak cannot be generated; above one, both the probability of an outbreak and its shape changes with increasing S/ρ ratio values. To simplify Kendall's arguments, we illustrate the waves generated at positions I, II, and III.

In wave I the susceptible population is only slightly above the threshold value. If an outbreak should occur in this zone, then it will have a low incidence and will be symmetrical in shape with only a modest concentration of cases in the peak period; as Figure 1.10 shows a Kendall wave I approximates that of the normal curve. In contrast, wave III is generated when the susceptible population is well above the threshold value. The consequent epidemic wave has a higher incidence, is strongly skewed towards the start, and is extremely peaked in shape with many cases concentrated into the peak period. Wave II occupies an intermediate position and is included to emphasize that the changing waveforms are examples from a continuum.

The spatial search for the kinds of waves predicted by the Kendall model has revealed some unexpected results. Andrew Gilg analysed the shape of waves generated by an outbreak of Newcastle disease in poultry populations in England and Wales.[37] Study of Gilg's maps suggest that Kendall

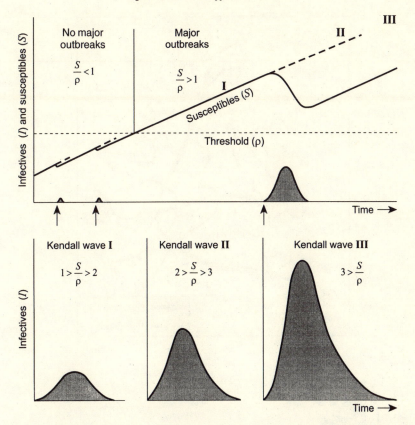

Fig. 1.10. Kendall model of the relationship between the shape of an epidemic wave and the susceptible population/threshold ratio (S/ρ). (*Above*) Growth of a susceptible population over time showing the effect of infections (see arrows). (*Below*) Three typical Kendall waves shown at locations I, II, and III on the upper graph.

type III waves are characteristic of the central areas near the start of an outbreak. As the epizootic spreads outwards, so the waveform evolved towards type II and eventually, on the far edge of the outbreak, to type I.

A generalization of Gilg's findings is given in Figure 1.11. This shows (A) in an idealized form the relation of the wave shape to the map of the overall outbreak and in (B) the waveform plotted in a space–time framework. In both diagrams there is an overlap between relative time as measured from the start of the outbreak and relative space as measured from the geographical origin of the outbreak.

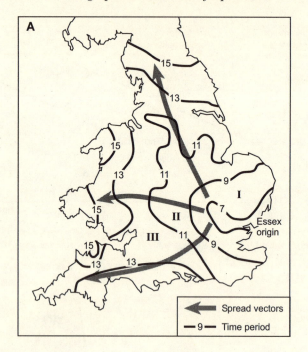

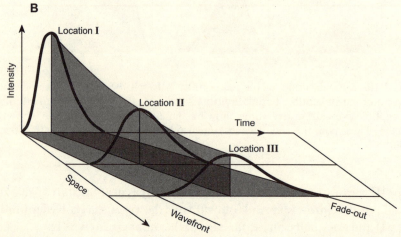

Fig. 1.11. Changes in the shape of epidemic waves with distance from the origin of an epidemic. (A) Gilg's findings on the spread of Newcastle disease epizootic in England and Wales. The epizootic started in Essex and the contours of spread are in 15-day steps (i.e. 11 = 165 days after the start). (B) Generalized wave model set in time and space: locations I, II, and III refer to the positions on the upper map. *Source*: Redrawn from Gilg, 1973, *op. cit.* [note 37] in Cliff and Haggett, 1988, *op. cit.* [note 20], Fig. 5.6B, p. 183.

If we relate the pattern to Kendall's original arguments in Figure 1.10, then we must assume that the S/ρ ratio is itself changing over space and time. This could occur in two ways, either by a reduction in the value of S, or by an increase in ρ, or by both acting in combination. A reduction in the susceptible population is plausible in terms of both the distribution of poultry farming in England and Wales and by the awareness of the outbreak stimulating farmers to take countermeasures in the form of both temporary isolation and, where available, by vaccination. Increases in ρ could theoretically occur either from an increase in the recovery coefficient (c) or a decrease in the diffusion coefficient (b). The efforts of veterinarians in protecting flocks is likely to force a reduced diffusion competence for the virus.

Epidemics as spatial diffusion processes

As is typical in many areas of scientific work, research was paralleled in different parts of the world. In Sweden, Hägerstrand's pioneering work on spatial diffusion was essentially concerned with wave transmission but set in a different context.[38] This was achieved through Monte Carlo simulation of spatial processes (i.e. by providing a well-defined copy or mimic of an observed historical sequence). In the years following its publication in 1953, his model was adapted to apply to regional cases as unlike as the spread of ghetto housing in American cities to early settlement of Polynesian islands.

Given the diffusion process being studied, there are three relevant questions that a geographer may wish to ask.[39] First, can we identify what is happening and why? From an accurate observation of a sequence of maps we may be able to identify the change mechanism and summarize our findings in terms of a *descriptive model* (see Figure 1.12).

Second, what will happen in the future? If our model can simulate the sequence of past conditions reasonably accurately, then we may be able to go on to say something about future conditions. This move from the known to the unknown is characteristic of a *predictive* model: the basic idea is summarized in the second part of Figure 1.12. We are familiar with this process in daily meteorological forecast maps on television or daily newspapers. If we have a series of slices through time, we search for the operator which appears—in terms of our model—to convert early maps of the past into later maps of the past and present. If we can find such an operator and if we can assume some continuity in this process (the heroic assumption that lies behind so much forecasting) then we can sketch in a future map. The error

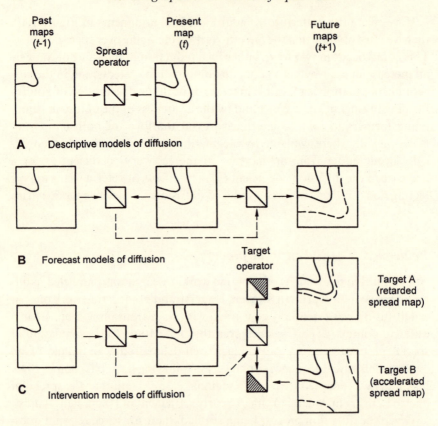

Fig. 1.12. Map sequences as a predictive device. Descriptive, predictive, and inter-dictive models of a spatial diffusion process. The creation of operators (e.g. mean information fields; MIFs) from a map sequence allows the forward projection of the map to a future time period.
Source: Haggett, 1990, *op. cit.* [note 2], Fig. 5.3, p. 103.

bands associated with each 'contour' will be given by the width of the lines on the map; the further forward in time, the wider those bands are likely to be.

But planners and decision-makers may want to alter the future, say, to accelerate or stop a diffusion wave. So our third question is: What will happen in the future if we intervene in some specified way? Models that try to accommodate this third order of complexity are termed *interdictive models*. I use this threefold distinction as a framework. We look at an application of the first group of models in the next lecture and reserve predictive and inter-dictive models until the last lecture.

2

Epidemics on Small Islands

Ever since Charles Darwin landed on the Galapagos Islands in September 1835, small islands have secured a special place in scientific history. In this lecture I try to show how the complex warfare of island invasion and island extinction observed by Darwin for plants and animals, also applies with equal force to the invasion and extinction of those microorganisms which cause many epidemic diseases. While case studies range from the familiar (influenza and the common cold) to the exotic (kuru and tsutsugamushi), I concentrate on the tracker diseases (measles) described in the opening lecture and confine myself to the islands Iceland and the Fijian group, which we have studied with most care. A continuing theme in both islands is the way in which technical development over the last 150 years, first in transport and then in vaccination, have fundamentally affected the ways in which epidemics circulate around the globe as diffusion waves.[1]

As Darwin argued, islands form natural laboratories in which processes can be observed which are too complex to track in the rapidly interacting city worlds that dominate the crowded continents. On small islands too, the level of observation is higher; as Panum observed in his Faeroes epidemic report: '. . . every contact with the inhabitants is known to every one and what is extraordinary, it is often recorded on the calendar'.[2]

Islands as natural epidemic laboratories

If we start with the conventional definition of an island as land surrounded by water then we have an immense galaxy of islands from which to choose. These range from the largest (the pancontinental Old World island of Africa–Europe–Asia) down through many orders of areal magnitude to the smallest rock. Size and frequency are inversely related. The smaller the

areal definition of an 'island' we are prepared to concede, the larger the estimated number in our presumed population. But since we are concerned
with epidemics, we require islands to be large enough to sustain (or to have
sustained in the past) human populations of sufficient size to act as the host
for epidemic-causing diseases.

Even within that subset of populated islands we have the further need
that islands have adequate medical records to follow in some detail the
progress of an epidemic. Thus although the world population of islands is
probably of the order of 10^{20} and that of populated islands 10^6, those with
adequate medical records is probably down to 10^2, a few hundred at most.
In *Island Epidemics* we have identified only eighty-four islands as having an
appropriate combination of demographic and epidemiological data to be
of great use for epidemic studies.[3]

We noted in the first lecture the fundamental work by Maurice Bartlett
on the relationship between epidemic wavelength and the population size
of British and American towns. A major difficulty in using such urban settlements in continental areas within which to study epidemic spread is that
models assume isolated geographical units that function as closed systems.
But in cities there is a constant leakage of population and disease into and
out of the area. These swash and backwash effects will be difficult, if not
impossible, to disentangle from the operation of the model process within
the area.

By choosing islands for study, leakage will be reduced to a minimum and,
importantly, these tend to occur at well-defined times when ships and/or
aircraft bond the islands to continental reservoir areas. With islands there
is a greater chance of population flux (whether by migration, journey to
work, tourism, or other passenger flows) being accurately recorded at ports
or airports. Accurate estimation of the degree of population turbulence in
terms of outflows and inflows plays an important part in any spatial epidemiological modelling.

It is therefore not surprising that island communities have had a central
place in biogeographical work since Charles Darwin.[4] They have also
played no small part in attempts to understand epidemics. Between
Panum's account of a measles epidemic in The Faeroes in 1846 and a modern virological account of the same disease in the Marshall Islands today
stands 150 years of study.[5] The isolation of islands and their small populations has meant that measles was introduced from time to time from
endemic continental areas. The sporadic pattern of waves that ensued
ranged widely over Bartlett's types B and C in Figure 1.8. Bartlett's work on

cities was paralleled by an early study by Francis Black of Yale University into measles outbreaks on islands.[6] Figure 2.1 gives his findings on the relationship between the degree of measles endemicity (vertical axis) and population size (horizontal axis) for 18 island communities. Only large islands, notably Hawaii with a total population of 550,000 at the time of Black's study, sustained continuous records of measles infection over the 16-year study period. The figure also indicates the estimated reporting rate obtained by comparing the number of cases reported with the birth rate and with serological tests. Of the five islands with rates in excess of 40 per cent, only one (Iceland) displayed a type B pattern; all the remainder were much smaller in population size and had irregular type C patterns.

Iceland as a laboratory

We have spent a considerable time over the last thirty years in studying epidemics on the island of Iceland. What are the reasons that lie behind this choice? The first reason for choosing Iceland is that it is exactly the right

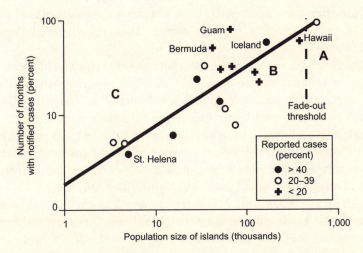

Fig. 2.1. Epidemic frequency and population size relationships for island communities. The Black model relating epidemic periodicity to population based on 18 islands. The vertical broken line indicates the endemicity threshold.
Source: Cliff *et al.* 1993, *op. cit.* [note 1], Fig. 1.9, p. 12.

size in population terms. It lies just below Bartlett's 250,000 threshold so that we can expect it to display Bartlett type B waves. The second most populous island in Black's sample, it is a large island (about the size of southern England or the state of Indiana) and is located just south of the Arctic Circle. The biggest single settlement is the capital, Reykjavik, which has been growing both in absolute population size and in its relative share of the Icelandic total. In 1901 its 6,700 inhabitants accounted for less than a tenth of the island's 78,000; by 1990, Reykjavik had swollen to 100,000 out of a total population of 250,000.

The second reason for Iceland is that it is doubly isolated. Given its mid-Atlantic location, it is: (a) remote from the virus reservoir areas of northern Europe and North America, and (b) individual settlements are isolated from each other. This secondary isolation is shown by the geographical distribution of Iceland's population in 1973 as shown in Figure 2.2, using the method of proportional circles. Iceland is the least-densely settled country in Europe. The harsh environment of the interior plateau with its icefields and tundra has restricted settlement to the peripheral lowlands (stippled in Figure 2.2). Given a deeply indented fjord coastline around much of the island, communities tend to be rather separate and remote from each other. Until the end of the Second World War when a round-the-island road was completed and local airfields built, most communication was by sea. It is therefore often more appropriate from an epidemiological point of view to think of Iceland as an archipelago of population islands rather than as a single cohesive unit.

A third reason for choice is the exceptional quality of the country's demographic and epidemiological records. These are among the most complete in the world. Records in reasonable detail go back to 1751 and through them it is possible to build up an accurate picture of Iceland's demographic and epidemiological history over the last 200 years. On the basis of birth rate and serological information, Black considered Iceland's data on measles case levels to have an accuracy of better than 40 per cent, substantially higher than in most other Western countries.

Fourth, these records are available on a very fine grid both in time and space. Both kinds of data are included in the one publication, *Heilbrigdisskyrslur*, the annual report of the state of public health in Iceland. First published in 1881, these records not only provide, on a monthly basis, figures for cases of infectious disease reported in each of some fifty medical districts (see Figure 2.3), but they also contain written accounts by local medical officers of the course of the disease in their own

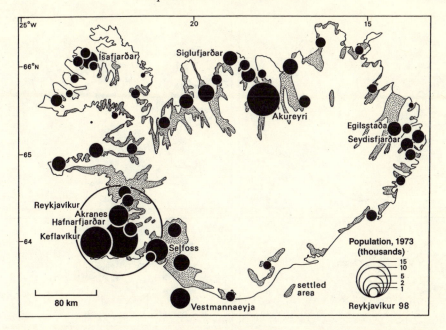

Fig. 2.2. Settlement distribution in Iceland. Population distribution in Iceland, 1973. Settlements are shown by proportionate circles and areas of cultivated land are stippled.
Source: Cliff and Haggett, 1988, *op. cit.* [note 1], Fig. 6.5B, p. 247.

district. These give details of the severity and spread of the various epidemics in each community and indicate, where known, the external source of the disease and how it diffused from village to village or even from farm to farm within each district. Not surprisingly, this microgeographical information was easier to obtain when most movements were by local boats. But with the coming of the motor car, aircraft, and tourism, the quality of this evidence deteriorates.

As with so many formats for data collecting, the names and boundaries of these fifty or so districts have been periodically reviewed during the 150 years since 1840. Major revisions were undertaken in 1875, 1899, 1907, 1932, and 1955. At each major boundary revision, names of medical districts were often changed, and new districts were created by dividing or amalgamating old districts in response to population changes. By detailed mapping, a series of consistent medical areas can be derived to permit spatial comparisons over time.

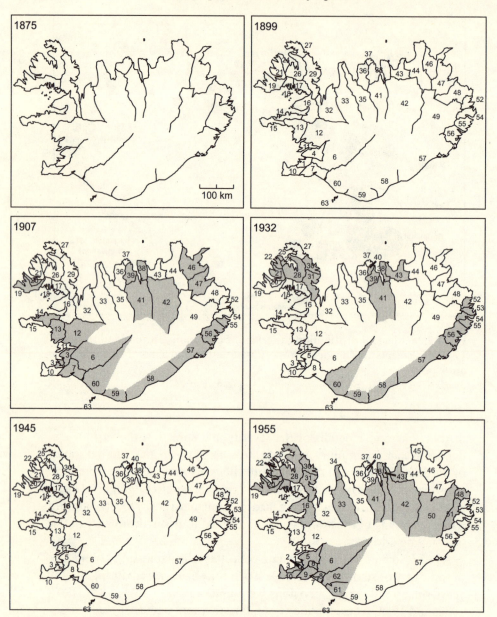

Fig. 2.3. Changes in the boundaries of recording districts for Icelandic medical data over an 80-year period. Base maps of medical districts at dates of major boundary revisions between 1875 and 1955. Each medical district is made up of the practice of one or more physicians. Boundary changes occur over time by splitting and amalgamating districts in response to population changes.

Source: Cliff and Haggett, 1988, *op. cit.* [note 1], Figs 2.11A–G, pp. 87–8.

A critical figure in establishing the disease recording system was the appointment of a new chief medical officer for the island, Jonas Jónsson, in 1895. Having studied in Panum's department at Copenhagen, he realized the importance of monitoring epidemics and reorganized the medical districts in 1899 and ensured accurate and regular reporting of infectious diseases. The picture of the diffusion of epidemics in Iceland presented here rests on the records of a handful of qualified men. Even by 1900, there were only 60 qualified doctors practising in Iceland. Jónsson has provided a vivid account of the work of the rural doctor in such districts, where the distance from the doctor's residence to an outlying farm might be as much as 112 kilometres:

Over this the doctor must make his way, often across mountains and uninhabited areas, or over desert sands and unbridged rivers. At the best such journeys are made on horseback, but in winter when even this method of travelling is impossible the doctor has often to go on foot, or on skis; in other places the journeys are made by boat across fjords and bays and even round promontories which project right out into the open oceans. At such times it may take the doctor the whole of a long day or even days to reach his patient and cost him and those who have to come to fetch him . . .[7]

But there were major differences between the towns and the sparsely settled districts.

A fifth reason for choosing Iceland for study was the major geographical changes taking place over the twentieth century. Iceland has experienced a population explosion since 1900 which reflects in the early years the declining mortality, especially among infants, and latterly, sustained high birth rates. This population has shown a major spatial reorganization, abandoning the more remote settlements and moving into the Reykjavik suburbs. Children's schooling in the rural areas has moved from a system of peripatetic teachers visiting farms on a rotational basis to one of children being concentrated into boarding schools. Easy accessibility by aircraft means that many older children now fly to and from the capital each week for their high school education.

Iceland's record of epidemic waves

We have noted earlier that Iceland has a unique epidemiological record. If we scan those records for the tracker disease (measles) it is possible to build up a month-by-month picture of the incidence of the disease. This shows that over a period of 150 years from 1840 to 1990 over 93,000 cases of measles were recorded. Over 99 per cent of these cases occurred in the form of nineteen distinct epidemic waves and these are identified by the Roman numerals in Table 2.1.

Table 2.1. Multiple epidemic waves of measles for Iceland over a 180-year period, 1800 to 1980

Wave	Date	Estimated deaths	Estimated cases	Duration (months)	Time gap between epidemics and geographical origin of virus (where known)
I	1846	2,026	na	6	Long
II	1882	1,700	na	6	Long
III	**1904**	**23**	**822**	**8**	**Long; Norway**
IV	1907–8	354	7,398	15	Medium; Denmark
V	1916–17	118	4,944	14	Long; The Faeroes, Norway
VI	1924–6	34	6,130	27	Long; Norway
VII	**1928–9**	**16**	**5,317**	**17**	**Short; Norway**
VIII	1936–7	60	8,408	14	Long; The Faeroes, England
IX	1943–4	18	7,155	18	Long; US or UK
X	1946–8	6	4,791	22	Medium; Denmark
XI	1950–2	11	6,645	26	Short
XII	1953	1	1,872	10	Short
XIII	**1954–5**	**4**	**7,787**	**21**	**Short**
XIV	1958–9	3	7,102	22	Medium
XV	1962–4	6	7,405	26	Medium; US
XVI	1966–8	4	6,152	22	Medium
XVII	1968–70	0	3,625	27	Short
XVIII	1972–4	0	3,953	16	Short
XIX	1977–8	0	3,223	15	Medium

Note. The three sample epidemic waves are in bold type.
'Long', gap of five years or more; 'Medium', three to five years; 'Short', less than two years; na, not applicable.
Source: Cliff *et al.* 1981, *op. cit.* [note 1], Table 4.1, p. 80.

The fact that the waves are separate and discrete with a virus-free window between each wave makes them much easier to analyse than the continuous and overlapping wave trains experienced in continental areas where the measles virus is endemic. Measles waves arrive discretely in both time and space in the island's settlements, and it is possible to follow the patterns of spread around the island. In Table 2.1, I have taken just three of the waves for detailed comment: waves III, VII, and XIII. Each illustrates the changing nature of the epidemics as the waves spread through the changing island population.

The first example (wave III) occurred during the summer of 1904.[8] The island had been free of measles since 1788 but, in 1846, it was affected by the same outbreak which ravaged The Faeroes (wave I). Another major measles epidemic came after a further interval of 36 years (wave II). The 1904 epidemic was the shortest and most spatially confined of the Icelandic waves analysed. In its seven months' duration, it spread only to the fishing villages and isolated farmsteads of north-west Iceland, and at no time were the two leading towns of Reykjavik and Akureyar affected.

The general course of the epidemic in north-west Iceland, from its commencement in April 1904 until its demise in November of that year, is outlined in Figure 2.4. Some 2,000 cases of the disease were recorded in a population of less than 11,000, an attack rate of nearly 20 per cent. As the inset histogram of number of cases against month of reporting shows, more than two-thirds of the cases occurred in the months of July and August, when the three doctors in the region were overwhelmed and noted in their journal that the epidemic was 'out of control'. The number of reported cases of measles by medical district is given using proportional circles. These indicate that the disease was focused on the main town in the region, Ísafjarðar. Of the 2,000 cases of measles which occurred in the region as a whole, about 1,500 were reported in Ísafjarðar alone, where new cases occurred at the rate of some twenty a day for two months and involved half the population of the area.

The doctors' accounts relate how measles arrived in the north-west fjords of Iceland in late April when a Norwegian whaling ship visited the whaling station in Hesteyrar. It brought the disease from its home port near Bergen in Norway where an epidemic was in progress. Despite attempts to contain the disease by quarantine, the crew members came into contact with the crew of a local shark-fishing vessel and both subsequently set sail for the town of Ísafjarðar and the whaling stations near the local farm and church of Eyri, taking the disease with them. Quarantine succeeded in eliminating secondary cases in the main town but failed to snuff out the disease at the whaling stations near Eyri.

Using local parish records, we have reconstructed the details of the course of the epidemic in the two contiguous parishes of Eyri and Ögur that lie on the southern side of the main fjord.[9] A brief lull in the spread of the epidemic was sustained until until 21 May when a Whitsuntide confirmation ceremony occurred at the church in Eyri. Most of the confirmed children were about 14 years old. The church was packed with the families and friends of the children. Many of the adults had been in contact with the

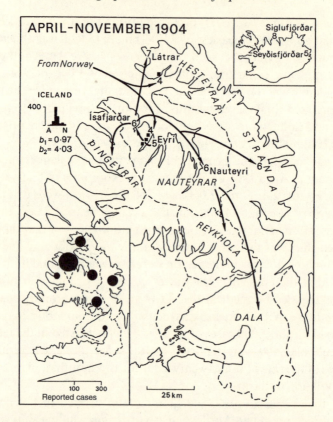

Fig. 2.4. Spread of wave III of the measles epidemic in north-west Iceland, 1904. Squares indicate whaling stations and vectors the inferred lines of epidemic spread based on reports of physicians in each parish. The histogram shows the reported number of measles cases each month, and numbers on the map (4, 5, 6) indicate the month in which measles cases were first recorded. The number of reported cases in each of the seven affected medical districts is shown (inset map) using proportional circles.

Source: Cliff *et al.* 1981, *op. cit.* [note 1], Fig. 4.3, p. 63.

crew members of the infected whaler and some of the crew members of the whaler itself were also present. Since, as noted in Table 2.1, the previous measles epidemic to affect the region was in 1882, many adults were susceptible as well as children. As a result of this mixture of contacts, the epidemic was reinvigorated on a grand scale in the days that followed. The critical role of the church ceremony in this process was recognized by the district's chief medical officer:

the spread of the disease was not out of hand until after Whitsun, 22 May; that day [*sic*] there was a service at Eyri, children were confirmed, and a large number of people were at the church. After that the disease spread very fast.[10]

The 1904 outbreak is of special interest as it underlines three factors of prime importance in understanding the spread process. First, it highlights the role of a single infected individual (the index case) whose spatial mobility plays a crucial part in starting an outbreak in a given community. Fishermen (or the crew of whalers in the 1904 outbreak) play a crucial role in the Icelandic case. Second, it indicates the importance of communal activities which bring susceptibles together at a critical period; the confirmation ceremony held at Eyri church on 22 May 1904 illustrates this point. Third, it describes the spatial countermeasures taken to control the spread of disease: placing ships in quarantine, isolating an infected farmstead, and putting patients into fever hospitals. In addition, the account gives us one of our few indications in the records of reporting errors. In Ísafjarðar, the doctor estimated that there were 1,500 cases; while the data record gives 310, a reporting rate of as low as 20 per cent, if the estimate is well founded.

The second wave I have chosen for comment (wave VII) occurred in the inter-war years.[11] By then Iceland's population had grown from 70,000 to over 100,000 and its capital city (Reykjavik) had tripled to 30,000. The seventh wave (shown in Figure 2.5A) came hard on the heels of the sixth, separated from it by a gap of only two years. It is not surprising to find, therefore, that this wave was both less intense and less prolonged than its predecessor. The total number of reported cases was 5,317 with sixteen deaths. The wave had a single peak in January 1929.

The pattern of diffusion through the Icelandic community was a distinctive one. The starting point in this wave was Siglufjarðar on the north coast, with an introduction from Norway into the fishing port. Within the next month, cases were being reported both from other districts in the same part of the north coast, and from Reykjavik and Vestmannoeyjar Islands in the south-west. Three separate subsystems can be identified thereafter: (i) continued spread from the original northern centre in the autumn of 1928 through the remaining northern districts; (ii) secondary spread from Reykjavik in the winter months of 1928–9 to other western and south-western districts; and (iii) spread between the eastern coastal districts, also in the winter months of 1928–9. The provenance of the third subsystem is

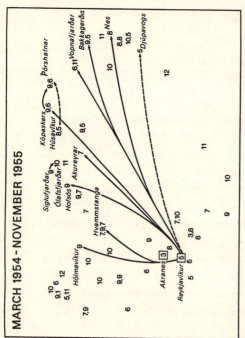

MARCH 1954 – NOVEMBER 1955

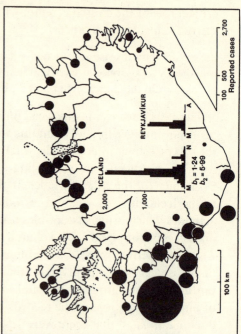

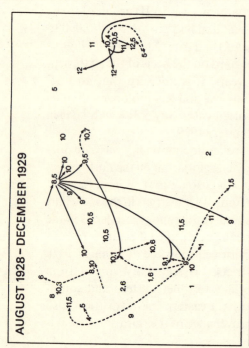

AUGUST 1928 – DECEMBER 1929

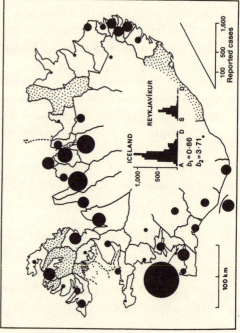

not clear, but two small fishing ports were reporting cases in October 1928, presumably stemming from the original Siglufjarðar introduction.

The doctors' reports lay stress on the role of boat crews in passing the disease between the coastal settlements. Even after the wave had passed, quarantine arrangements for ships remained in force. Thus, the Siglufjarðar report for 1932 relates how boats were loaded under the supervision of the police to prevent infected crew members bringing measles ashore. Other reports describes the work of the local 'committee for the prevention of disease' in placing quarantine orders on ships with crews suspected of having the disease.[12]

I have added a third wave (wave XIII) to typify the period after American and British occupation during the Second World War. Not only was the death rate by now negligible but the capital city Reykjavik now dominates the spread pattern. This reflects the growth of air transport and the relative decrease in local shipping as a means of passenger transport. Together, the three waves show the increasing scale of the epidemics. Wave III was local, wave VII had a regional structure, while wave XIII was essentially island-wide. The pattern of vectors also shows the changing pattern of spread. The first two waves are dominated by ship movements, the third by direct aircraft routes from the capital city.

Genereralizations and wave sequences

While the study of individual waves gives some idea of the changing character of epidemics, we need to use all the waves if we are to be able to build general models. A clear indication of this is given in Figure 2.6. This shows the spacing of the waves becoming shorter as Iceland's population rises and contacts with the outside world become closer. The average gap between waves in the period from 1896 to 1945 was more than five years; from 1946

Fig. 2.5. Spread of waves VII and XIII of the measles epidemic in Iceland. (A) Wave VII, August 1928 to December 1929. (B) Wave XIII, March 1954 to November 1955. In both cases the upper maps show the known vectors of spread based on physician's reports from each district. The lower map shows the distribution of reported cases using proportional circles with medical districts not recording cases stippled. Separate histograms show the time profile of cases both for Iceland as a whole and for Reykjavik.

Source: Cliff *et al.* 1981, *op. cit.* [note 1], Figs 4.7, 4.13, pp. 73, 84.

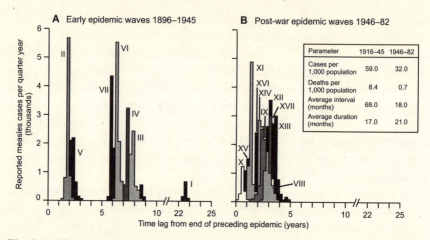

Fig. 2.6. Time intervals between measles epidemic waves in Iceland. The diagram shows contrasts between the earlier (widely spaced) and later (closely spaced) measles epidemics over the period 1896–1982.

Source: Cliff and Haggett, 1988, *op. cit.* [note 1], Fig. 6.5J, p. 253.

to 1982 it had fallen to a year and a half.

The simplest way to show changes in the geographical distribution of measles over that time is to arrange maps in sequence, like individual stills from a moving picture. By itself, the monthly map sequence for a single epidemic wave is hard to interpret. But when several sequences of such epidemic maps are placed together, we can with reasonable luck begin to pick out the main threads of the ways in which measles spreads. Figure 2.7 generalizes the results for eight epidemic waves that hit Iceland from 1945 to 1970. The average time lag (in months) from the start of an epidemic before each settlement was reached is given.

The diagram reveals four phases. (1) An *originating phase* (with lags of less than three months) when measles is confined to the capital city, Reykjavik. (2) A *localized spread phase* (third to fifth months) when measles spreads locally in a neighbourhood fashion to communities around the

Fig. 2.7. Average lag structure for measles epidemics in Iceland, 1945–70. Mean lag time in months from the start of all measles epidemics occurring between 1945 and 1970 (waves X–XVII) before medical districts were reached. Stipple indicates districts shown as reached on earlier maps. Area of circles proportional to population size of districts in 1970.

Source: Cliff and Haggett, 1988, *op. cit.* [note 1], Fig. 5.8B, p. 189.

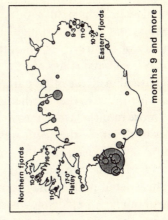

less than 3 months

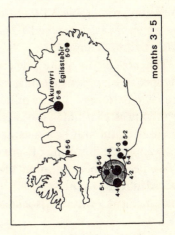

months 9 and more

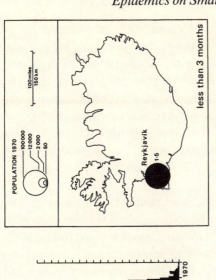

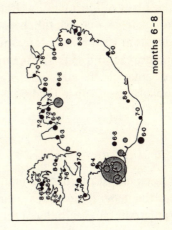

months 6 - 8

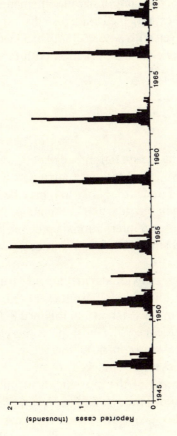

months 3 - 5

capital (e.g. Selfoss in south-east Iceland with an averaged lag of 5.3 months) and by longer-range diffusion to regional centres (e.g. the northern city of Akureyri, 5.8 months) in the northern half of the island. (3) *Generalized spread* (lags of six to eight months) results in measles becoming established in all the other parts of the island, with the exception of the two zones identified in the fourth phase. (4) A *remote outliers phase* (lags of nine and more months) when measles is delayed in getting into two inaccessible parts of Iceland—the north-west fjords (e.g. the remote fishing villages of Djúpavikur, sixteen months, and Flateyjar on Flatey Island, seventeen months) and the eastern fjords.

This spatial pattern may seem to lend support to the idea of a simple spread model for epidemics, namely, moving down the hierarchy of urban size and the contagious spread of epidemic waves moving out from the initial centres of introduction. But such an interpretation is complicated by the high inverse correlation between population size of settlements and distance from Reykjavik. Many smaller settlements are also distant from the source areas, and so it is difficult, even with powerful statistical techniques, to disentangle the size and distance effects. In practice, both factors were probably operating in tandem.

Iceland as a graph

Another useful approach to mapping the spatial spread of an epidemic is to treat the geographical area over which spread is occurring as a graph.[13] Graphs consist essentially of sets of nodes and links and have been widely used in geographical analysis, both in physical and in human geography.[14] We know from Table 2.1 that the known international epidemic pathways by which measles reached Iceland since 1896 show a strong bonding to Norway and Denmark.

The movements of the index cases (individuals carrying the virus) can be mapped in a similar way, on a monthly basis, to permit examination of the seasonal pattern of disease diffusion. This is shown in Figure 2.8. In northern temperate latitudes, measles epidemics, like most infectious childhood diseases, commonly have a winter peak. In contrast, however, in Iceland the May and June flowering of epidemics is most marked.

We need to bear in mind that most of the movements of index cases recorded relate to the pre-1945 period. Thus, we find in *Heilbrigdisskyrslur* that medical officers associate this peak with two facts, namely, the improved communications between districts after the spring thaw, which permitted greater spatial mobility, and the occurrence of the major communal activity of haymaking in June. In fact, the seasonal pattern is more subtle than the monthly maps imply for, in the pre-1939 period, there were two epidemic peaks of roughly equal intensity—the summer one and also the midwinter one found in other Western countries. Nor has the pattern been stable through time. In the post-1945 period, the summer peak has gradually disappeared as Iceland's isolation from the rest of the world has been eliminated by airline travel.

Iceland: predicting epidemic measles waves

The detail and clarity of evidence for epidemic movements in Iceland have encouraged experiments in model building. Andrew Cliff has explored a number of different predictive models for epidemic waves shown by Icelandic measles data using the period between 1945 and 1970.[15] Table 2.2 summarizes the models developed and the medical districts within Iceland to which they have been applied. The main strengths and weaknesses of each are also listed. On the basis of this table, several main conclusions can be reached:

Table 2.2. Forecasting models applied to measles epidemics in Iceland

Model	Application format		Principal data inputs		Temporal parameter structure		Comments
	Single region	Multi-region	S	I	Fixed	Varying	
Hamer–Soper	To all medical districts		×	×	×		Good at forecasting epidemic recurrence years ahead; average to poor on estimating epidemic size.
Chain binomial	Reykjavik		×	×	×	×	Initial one-month lag effect; able to lock on to course of epidemic; reasonable estimates of epidemic size.

Table 2.2. *cont.*

Model	Application format		Principal data inputs		Temporal parameter structure		Comments
	Single region	Multi-region	S	I	Fixed	Varying	
Autoregressive	Reykjavik			×	×		One-month lag effect; consistently misses epidemic starts; reasonable estimates of epidemic size.
	Reykjavik			×		×	Initial one-month lag effect; adapts to changing phase characteristics; over-estimates epidemic size.
GLIM		North-west Iceland (*n* = 5)	×	×	×		As fixed in time parameter chain binomial.
Kalman filter	Reykjavik			×		×	Initial lag effect, but locks on to epidemic course at expense of overestimates of epidemic size.
Bayesian entropy	Reykjavik		×	×		×	Despite separate epidemic/no epidemic models, one-month lag effect in predicting epidemic curve; probability forecasts of epidemic/no epidemic good; model switches in correct month.
Simultaneous equation		Multi-region chains		×	×		Areas studied as causal chains; good phase characteristics guaranteed by formulation of model; non-registration or serial interval and data recording interval restricts use in forecasting.
Logistic transformation		Multi-region chains		×	×		Good probability forecasts of epidemic/ no epidemic states; slow state switching.

Note. GLIM, general linear model.
Source: Cliff *et al.* 1993, *op. cit.* [note 1], Table 15.8, p. 410.

1. No predictive model produces accurate projections of both the recurrence times and size of epidemics. Improvement in one tends to be bought at the expense of the other. Thus, if a model is devised that will forecast wave recurrence acceptably, then epidemic size is overestimated. To forecast inter-epidemic times accurately, it is generally necessary to tune the model to be sensitive to changes signalling the approach of an epidemic, with the result that it is too sensitive and tends to overshoot on forecast severity when the epidemic is in progress.

2. Models that are based only on the size of the infective population in previous time periods consistently fail to detect the approach of an epidemic. Instead, they provide reasonable estimates of cases reported, but lagged in time. They accurately shadow rather than extrapolate the epidemic behaviour.

3. Models with parameters fixed through time have a tendency to smooth through epidemic highs and lows because they are unable to adapt to changes between the build-up and fade-out phases. Time-varying parameter models are better at avoiding this problem.

4. Epidemic recurrences can be reasonably anticipated only by incorporating information on the size of the susceptible population and/or properly identifying the lead–lag structure among medical districts for disease transmission. Addition of spatial interaction information markedly improved our ability to forecast recurrences in lagging areas. Information on susceptible population levels also serves to prime a model to the possibility of a recurrence, as is made clear by the various threshold theorems. Models based on susceptible populations, but which are single- rather than multi-region, tend to miss the start of epidemics but rapidly lock on to the course of an epidemic thereafter. Models that are dominated by spatial transmission information, at the expense of information on the level of the susceptible population in the study region, produce estimates of epidemic size which reflect the course of the epidemic in the triggering regions rather than in the study region.

5. Probabilistic process models enhance our understanding of disease transmission across geographical space and increase the chances of devising time series models appropriate to the task of forecasting.

Cliff's conclusions highlight the fact that naive predictive models produce poor results. However, the Icelandic results indicate clearly the gains

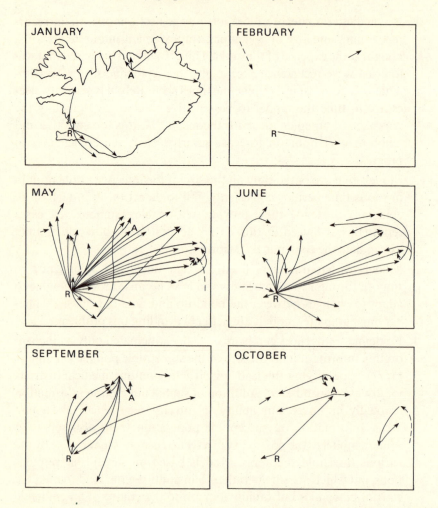

Fig. 2.8. Seasonal distribution of measles contacts in Iceland. The map shows on a month-by-month basis all recorded links over the 80-year period 1896–1975. Links are drawn directly between pairs of places but in practice, links would often have been round the coast by local shipping. The outline of Iceland is shown for January but omitted for clarity for other months. The location of the two main settlements Reykjavik (R) and Akureyri (A) are shown.

Source: Cliff *et al.* 1981, *op. cit.* [note 1], Fig. 5.24, p. 128.

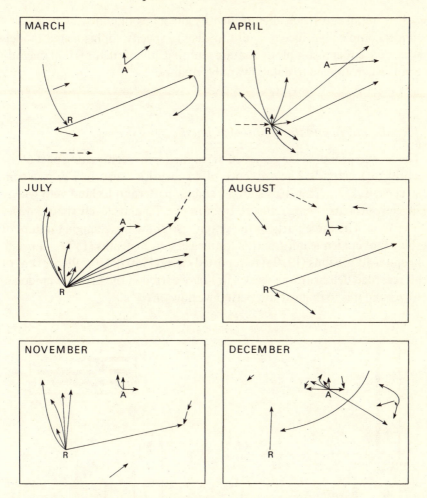

to be made for each extra element of complexity added to our models, namely: (a) time-varying parameters to handle the non-stationary nature of within-epidemic structure, particularly the fundamentally different character of the build-up and fade-out phases of an epidemic; (b) separate models for epidemic and inter-epidemic episodes to recognize the fundamentally different character of these periods; (c) spatial lead–lag information to improve our ability to forecast epidemic recurrences and to understand the transmission of disease between areas; and (d) incorporation of data on the susceptible population level to improve estimates both of epidemic size and likely recurrence intervals.

It is important to be able to identify the gains obtained by increasing model complexity, since it is all too easy to specify sophisticated models which are either insoluble or contain multiplicative structures that magnify errors when applied to data of variable quality.

Measles in Fiji and the south-west Pacific

The second island we have studied at length lies in a contrasting part of the world to Iceland.[16] The sequence of 14 measles waves in Fiji is reconstructed in Figure 2.9. This shows a similar pattern to Iceland with widely spaced outbreaks being replaced by more regular and closely spaced waves over time. The three parts of the graph also show the changing nature of sources of epidemic information for measles: from deaths (1870–) through hospital admissions (1910–) to reported cases (1940–). Over 99 per cent of all recorded deaths from measles in Fiji occurred in the first two epidemic waves and it is to that earliest period we now turn.

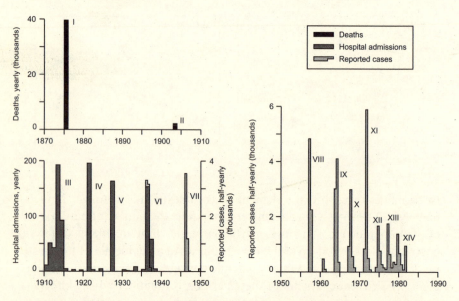

Fig. 2.9. Sequence of measles epidemic waves in Fiji, 1875–1982. Data used to identify wave change over time. Mortality and hospital admissions data are annual; reported cases are half-yearly.
Source: Cliff and Haggett, 1985, *op. cit.* [note 1], Fig. 6, p. 38.

The spread of measles from Australia and New Zealand into the rest of the south-west Pacific appears to have occurred in four distinct surges (see Figure 2.10): into New Zealand in 1835, into the Tahiti and Cook Islands in 1854, into Fiji, the Solomons, and New Guinea in 1875, and into Tonga and Samoa in 1893. By 1910, measles had been introduced at least once into the major islands of the most-populated island groups of the Pacific, and the disease was by then endemic in two new bridgehead areas—Australia and the west coast of North America. But there were many smaller islands that remained free of the disease until this century. For example, several of the Aleutian Islands were not affected until after 1900 and Rotuma, north of the main Fijian groups, was not reached until 1911.

The existence of endemic reservoir areas for the measles virus in populations around the edge of the Pacific has obviously changed over time. If we accept Bartlett's estimate (see Lecture 1) that a city with an unvaccinated population of about one quarter of a million can serve as a reservoir then, in 1800, we could only have found such concentrations around the north-west rim of the Pacific in the Japanese and Chinese populations. The dense rural population on the island of Java would also have served as such a base within the Indonesian archipelago. As the nineteenth century progressed, so the number of cities crossing the threshold increased: from four by 1850, to twelve by 1900, and to nearly forty by 1950.

The Fijian outbreak of 1875

The first measles wave to strike the Fijian group was so disastrous in its impact and so rich in morbid detail that it forms a benchmark in Pacific history. It has been described in many accounts. One of the fullest is given in *Report of the Commission Appointed to Inquire into the Decrease of the Native Population* published in Suva by the Colony of Fiji in 1896.[17]

The story has often been told and may well have gained something in the telling. The official report on the decrease of the native population published eleven years after the end of the epidemic was concerned more with causes and cures than with detailed reconstruction of spread. Likewise, much of the official correspondence nearer in time to the epidemic was preoccupied with avoiding blame rather than providing an accurate epidemiological description.

A major epidemic involving a specific virus is an event which would normally generate a rich vein of medical material. In the first Fijian wave, such

A.1800

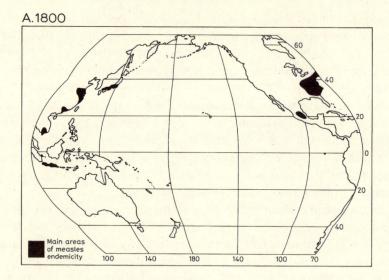

C.1851-60

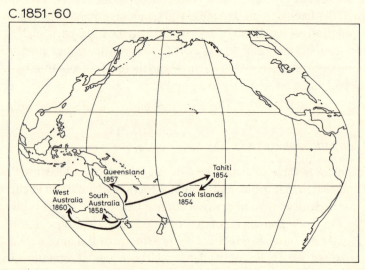

Fig. 2.10. Generalized directions of measles spread in the Pacific Basin since 1800. Dates of first introduction are shown where known. (A) Areas of measles endemicity, 1800. (B) Spread, 1801–50. (C) Spread, 1851–60. (D) Spread, 1861–1910. (Note that information for spread in South-East Asia is lacking.)

Source: Cliff and Haggett, 1985, *op. cit.* [note 1], Fig. 3, pp. 22–3.

B.1801-50

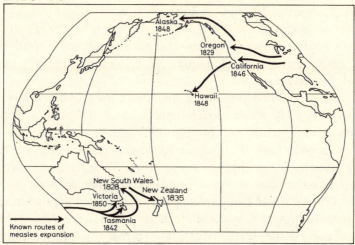

D.1861-1910

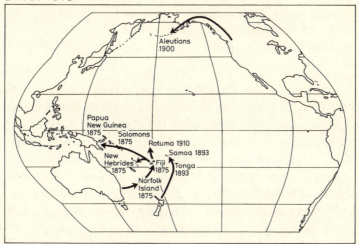

material is almost entirely absent; the arrival of measles immediately pre-dated the change from native rule to colonial status which was to bring a measure of Western medical organization to the islands. At the height of the crisis, events were being coped with by just two medical officers, assisted on occasions by a ship's surgeon from a visiting vessel.[18] How far the deaths that followed the introduction of measles were due to measles and its immediate sequelae and how far to associated diseases or malnutrition will never be known unambiguously. Even when William McGregor took up his duties as the first of Fiji's chief medical officers in June 1875 (at the very end of the epidemic) there were only four medical officers to serve 150,000 people scattered over 1.3 million square kilometres on 300 islands, of which less than 100 were inhabited.[19]

For evidence we are dependent on surviving correspondence, newspaper accounts, and, above all, on mission papers. The first Wesleyan missionaries landed in eastern Fiji from Tonga in 1835 and the first Catholic mission was established in 1844. By 1875, the Wesleyans had an extensive chain of mission stations throughout the Fijian group. With its district organization, its regular minutes and reports, and its annual register of members, the records of the Wesleyan church were particularly suited to provide an indirect source of information on the mortality which accompanied this first great onslaught of the measles virus.[20] They provided a summed estimate of the loss of life in the first half of 1875. Whether due directly or indirectly to measles or not, the proportions of 'missing' names on the Wesleyan church registers when the roll was called at the end of 1875 is the most pertinent measure of local variations in mortality.

The first few links in the infectious chain can be plotted with some accuracy (see Figure 2.11).

1. The arrival of HMS *Dido* at the Fijian capital, Levuka, on the east coast of Ovalau Island on 12 January 1875 was the opening move. The royal party on board consisted of the Fijian chief (Cakobau) and his two sons, who were returning to Fiji after a state visit to New South Wales as guests of its governor. The Fijian chief had contracted measles while in Sydney but had long since ceased to be infectious. The vessel had left Sydney on 21 December, but the two sons went down with measles during the twenty-two-day voyage via Norfolk Island. On arrival at Levuka, no yellow flag (indicating an infection on board) was flown by the ship and, although the disease was reported to the shore by the ship's surgeon, no attempt was made to stop either the welcoming group coming on board or the royal party disembarking. The status of the passengers, political sensitivity over cession to the British Crown, the lack of formal quarantine arrangements in Fiji, and a complete underestimate of the likely impact of measles in a 'virgin soil' (i.e. previously unexposed) community, all played some part in this oversight.

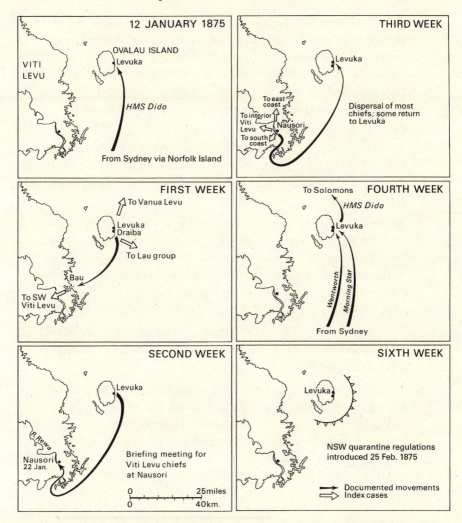

Fig. 2.11. Fiji, 1875: elements in the spread of measles in the first six weeks of the epidemic. Arrows indicate direction of spread where this can be documented. *Source*: Cliff and Haggett, 1985, *op. cit.* [note 1], Fig. 5, p. 32.

2. Over the course of the next ten days Cakobau entertained chiefs from Ovalau and the other islands at his homes at Draiba near Levuka and on the island of Bau (Figure 2.11, centre left). An unusual number of high-ranking visitors from the outer Fijian islands travelled over long distances both to welcome the senior chief home and to hear more about Fiji's new status as a British colony.

3. In the following week a major gathering of the chiefs of the hill tribes of the main island of Viti Levu was arranged at Nausori on the Rewa River (Figure 2.11, lower left). About 800 appear to have gathered there on 22 January to learn at first hand from the Ovalau chiefs what cession meant.

4. At the conclusion of the Nausori meeting, most hill chiefs went back to their homes, but a small party of five of the most prominent (together with their groups) returned to Levuka to inspect the visiting warship (Figure 2.11, upper right). By then Levuka was into its second generation of measles cases and the number reporting sick was multiplying alarmingly. All five chiefs were to die by the next month.

5. The chiefs returned to their different parts of Viti Levu and HMS *Dido* sailed (in early February 1875) for Malekula in Vanuatu (the Solomons); see Figure 2.11, centre right.

6. To make the epidemic bridgehead established by the *Dido* doubly secure, the *Wentworth* arrived at Levuka from Sydney on 26 January, followed by the *Northern Star* in early February (Figure 2.11, centre right). Both ships carried active measles cases and, again, the infected passengers were allowed to land. It was to be 25 February before the quarantine laws of New South Wales were proclaimed in Fiji (Figure 2.11, lower right).

It is important to note that we have direct medical evidence only of the two infected royal sons in link (1) and the passengers in (6). That the virus was passed on during links (2)–(5) is an inescapable inference from the news that began to come in from all parts of the island group. The unfortunate *Dido* returned workers from Fiji to the Solomons and Corney attributed the subsequent measles epidemic there and in the New Hebrides to this link.[21]

While the speed of spread and intensity of impact is striking, it is hard to imagine a diffusion hierarchy more calculated to accelerate spread of the virus. At each point in the linkage pattern of Figure 2.11, we have an

unusual number of people who have come long distances, who meet in close concourse, and who then return to all parts of Fiji where they are met in their turn by local groups. The news about cession to Great Britain and the advent of the measles virus appear to have spread hand-in-hand down through the layers of the social hierarchy.

The timing of the geographical spread of measles from then on is only partly known. Ovalau and Bau were clearly infected in early January and the returning chiefs could have spread the virus into most districts by late January or early February. Reports from mid February confirmed the presence of measles in all provinces and 'by the early or middle part of March disastrous news came in from all quarters'.[22] The peak incidence of the disease was reached in the original centres towards the end of March and, in most other places, by early April. The remote eastward group, the Lau Islands, was last affected but measles appears to have spread, albeit at a slower rate, into all the major islands including Ono-i-Lau, 390 km from Levuka.

The end of the epidemic showed a similar spatial pattern of slow withdrawal with those first affected also recovering first. By the middle of April, good news began to appear in the Fijian press. By early May, the disease had died out in Ovalau and the more accessible parts of Viti Levu; by early June, it had disappeared from the whole island group. Ironically, by the time the colony's first chief medical officer arrived, the epidemic was over.

The demographic impact of the epidemic

The full demographic impact of measles on the Fijian population will never be known. The 'unrecorded' pre-1881 census, taken in 1879, gave a count of 108,924 Fijians plus an estimated 3,000 working on European plantations. The official estimate of the population of the Fijian group at the start of 1875 was 150,000 and the final estimate of deaths was 'not less than 40,000'.[23] This would give a death rate of around 27 per cent.

The origins of these official estimates are obscure and recent research has tended to question them. Norma McArthur in her *Island Populations of the Pacific*, after an extended review of the evidence, suggests that a total pre-epidemic population nearer to 135,000 (and certainly not over 140,000) is more likely.[24] If we then apply the same number of deaths as in the official estimates, the death rate rises to 30 per cent; but if we retain the same rate (27 per cent) and apply it to the smaller population, total deaths would fall

to 36,000. In the light of current knowledge of measles death rates in virgin soil populations, a death rate of 1 : 5 rather than 1 : 4 would be a more realistic assessment. If we combine this with McArthur's lower population estimate, the loss of life might have been as low as 27,000. The real figure probably lies somewhere between this lower bound and the official estimate of 40,000, not the *Times* figure of 50,000.[25]

But whatever estimates are taken, the death rate from the initial Fijian measles outbreak was of awful proportions. Even conservative guesses suggest that the losses in the first half of 1875 amounted to more than one in five of the Fijian native population.

The demographic impact of measles outbreaks on virgin soil populations has been widely studied.[26] From Panum's detailed figures, a death rate from measles of 23 per 1,000 is indicated. This was about four times the normal death rate in The Faeroes at that period and was mainly made up of excess mortality in the infant and 50- to 70-year age group.

As Table 2.3 indicates, the death rate from measles outbreaks in virgin soil populations is high. But the Fijian mortality appears to be so much higher in comparison with most other measles epidemics, even in virgin

Table 2.3. Estimated death rates from measles in isolated populations

Year	Location	Years since previous outbreak	No. of deaths on which rate estimated	Estimated death rate per thousand
1846	The Faeroes	65 (1781)	180	23
1846	Iceland	58 (1788)	107	39
1860–1	Western Australia	57	57	2
1875	Fiji	'Virgin soil'	c. 40,000	250
1882	Iceland	36 (1846)	250	45
1900	Alaska	West Alaska: 'virgin soil' Aleutians: ?25 (1875)	1,000+	c. 250
1903	Fiji	28 (1875)	c. 2000	?
1952+	The Americas	'Virgin soil' Isolated Indian tribes	?	30–270

Note. 'Virgin soil' = previously unexposed.
Source: Cliff and Haggett, 1985, *op. cit.* [note 1], Table 4, p. 54. The original table includes for each epidemic a detailed list of sources on which the figures are based.

soil, that some special explanation might need to be investigated. Three factors may well have multiplied up the deaths.

First, is the sheer scale of the outbreak. It seems likely that more than 100,000 people were affected by the virus, all within a span of six months. The breakdown of normal life in terms of food gathering and hygiene, plus the contagious fear of a new affliction whose duration and outcome was unknown, all exacerbated the effects. The prevalence of dysentery and the lack of care for the very young, the very old, and the sick, are all documented. Even those who were medically fit were so psychologically overcome by the scale of the disaster that they ceased to care for their own health.

A second factor was undoubtedly due to the 'treatment' exacerbating the initial problem. The 1896 commission reported that:

One correspondent relates that during the epidemic of measles, in 1875, natives persisted in trying to cool their fevered bodies by lying for hours in the water and in damp places, in spite of strict orders to the contrary. A child heated with fever is plunged in cold water or set naked in the breeze or damp grass to cool its body. Such ignorance is to be expected among a people lately confronted with diseases unknown to their fathers and whose knowledge of diseases is solely derived from empiricism.[27]

How widespread the practice was is unknown, but it was still being resorted to sixteen years later during the 1891–2 influenza epidemic. Where normal rest and nursing was available, the death rate was much lower. In the Nadi area of Viti Levu, the local Justice of the Peace circulated pamphlets on *What to Do and What to Avoid if Taken with the Disease* to head teachers and village chiefs. He reported only a dozen deaths in eight villages with a population of more than 2,000 and suggested similarly low figures were typical among the imported labourers who were also following standard instructions.[28]

The third factor was the overlap of the measles epidemic with an unusually severe hurricane season. This may well have intensified the general health problems in the early part of 1875. It seems evident that in the early months of 1875, little would go right for the Fijian people.

Indian migration to Fiji

Despite the 1875 problems, the period following the 1875 outbreak was one of economic growth in Fiji. Like Trinidad in the West Indies and Mauritius in the Indian Ocean, Fiji's climate and soil were ideally suited to growing

sugar cane, a crop for which European demand was soaring. Unfortunately, all three islands were short of suitable local labour. In the post-slavery period, the import of slave plantation labour from Africa was banned. A new source area (India) and a new form of bondage (indentured labour) provided a solution.

The history of Indian migration to Fiji up to the end of indenture in 1920 has been extensively studied by Gillion.[29] Over the thirty-seven years between 1879 and 1916, Indian immigrant ships made eighty-seven voyages to Fiji carrying nearly 61,000 indentured emigrants. The health and welfare of the immigrants were the responsibility of the surgeon-superintendent who accompanied each ship and whose report was incorporated into the Annual Reports on Indian Immigration published regularly as Official Papers of Fiji's Legislative Council.

From these reports, the presence of measles can be detected and linked to various Fijian epidemics. For example, the 1911 Report records 'many cases of measles breaking out' on the *Mutlah*, and 'an epidemic of measles of a mild type' on the *Sutlej*, but the other three voyages in the second half of the year were free of that disease.[30]

Since measles was endemic in much of India, it is not surprising that cases should be recorded despite checks in the camps at Calcutta and Madras, the two exit ports, before embarkation. Close study of data in the Fijian Annual Reports, indicates that measles was detected on board during thirty-one of the voyages. These are broadly distributed over time and reflect the 1 : 3 probability of a ship being infected on departure from India. The distribution of voyages where vessels were still infected with measles on arrival in Fiji is shown in Figure 2.12. These amounted to only eleven, indicating that measles did not survive in two-thirds of the voyages in which it was known to be present at the outset. Such voyages all occurred late in the emigration period. The explosive growth of a measles epidemic is illustrated by a subsequent voyage of the *Fazilka* in 1903. In this instance, two cases were reported in the first half of the voyage, eight in the second half, but 130 in the quarantine depot at Nukulau Island within ten days of landing.

Some of the factors behind this distribution of infected arrivals are shown in Figure 2.12. This plots each voyage in terms of its length in days (horizontal axis) and the size of the vessel (vertical axis). Clearly, an important distinction must be drawn between the sailing ships used between 1879 and 1904 and the steamships used between 1884 and 1916.

Sailing ships followed the route south of Australia and took about seventy days for the voyage. Steamships came through the Torres Strait north

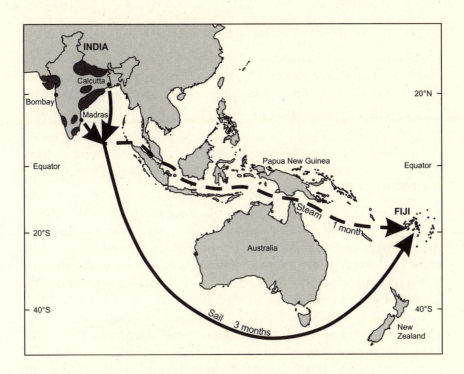

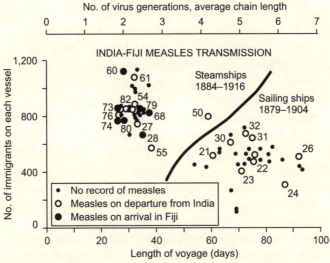

Fig. 2.12. The impact of changes in transport on disease introduction. (*Above*) Routes of vessels carrying indentured immigrants between India and Fiji, 1879–1916, categorized by sailing ships and steamships. (*Below*) Plot of voyages by length of voyage in days (this is also translated into 14-day measles virus generations on the upper axis), by type of vessel (sail or steam), and by measles status (see key). Numbers indicate the sequence of voyages.

Source: Cliff and Haggett, 1985, *op. cit.* [note 1], Fig. 12, p. 63.

of Australia and halved the sailing ship times; they were also able to carry a larger number of immigrants. The threat of speeding the introduction of measles into Fiji by using the faster and larger steamships was considered by the medical officers on the ships, but they did not rate the risk as critical:

So far as can be judged as yet the introduction of immigrants by steamers has not had a prejudicial effect on their health, though it increases the chance of introducing diseases of a severe type into the colony and renders more likely the necessity of imposing quarantine.[31]

For the virus to persist in a steamship voyage it needed only two, or at most three, generations of measles attacks, assuming an average link length of fourteen days. On the sailing ships, virus persistence required five or six generations, and this within a slightly smaller population of passengers. It is not surprising that it was the steamships, rather than sailing ships, that brought measles from India.

Even among those ships arriving with measles on board, the chance of starting an epidemic on Fiji itself was quite small. One factor was that all immigrant ships were quarantined and inspected on arrival. The first quarantine station was established on Yamuca Island between Ovalau and Moturiki islands, and was used by the first immigrants from the *Leonidas*. With the shift of the Fijian capital to Suva, the quarantine station was moved to the island of Nukulau on the reef about 10 km east of Suva harbour. Immigrants were usually detained for a fourteen-day period before being delivered to the plantation areas.

A second factor limiting the spread of measles was that an infected migrant (usually a child) would be joining an Indian plantation community in which most Indians would have a natural immunity due to previous infection. The proportion of susceptibles in the locally infected community was therefore low.

The Pacific represented the last major frontier for the measles virus. At the beginning of the nineteenth century, the populations of Australasia and the islands of the south-west Pacific were unfamiliar with the disease. By the outbreak of the First World War, it had swept throughout the area, and major reservoirs of measles virus had become established there. By the first decade of the twentieth century, the period of massive all-age loss of life from measles (as in the Fiji disaster) was over and the sickness had become largely confined to children.

In their different ways, both Iceland and Fiji show the means by which islands allow unique insights into the movement of epidemic waves. Their isolation and the ability to log the critical movements of index cases give them an advantage as windows on a complex spread process. In the next lecture we open those windows in both a spatial and temporal sense and go on to examine disease emergence at the global scale over a much longer time span.

3

Global Origins and Dispersals

The closing decade of the twentieth century saw a curious paradox in the history of epidemic diseases.[1] On the one hand, we observed a steady fall in mortality from the classic infectious diseases, culminating in the 1970s in the global eradication of one of the world's most dreaded killers (small-pox). This was the decade in which W. H. Stewart, the Surgeon-General told the United States Congress that it was '. . . time to close the book on infectious diseases, declare the war against pestilence won, and shift national resources to such chronic problems as cancer and heart disease'.[2] In like vein, biologist John Cairns could write in 1975 that '. . . the western world has virtually eliminated death due to infectious disease'.[3]

On the other hand, within the same decade a new pandemic disease, AIDS, was emerging that seemed likely to kill in the next two decades as many victims as had smallpox in all the previous decades of the twentieth century. Nor was AIDS alone. To it we can add such new diseases as legion-naires' disease, Lyme disease, and toxic shock syndrome, plus the out-breaks of African tropical diseases that occasionally erupt into middle-latitude consciousness, caused by the Lassa, Marburg, and Ebola viruses (see Table 3.1). The re-emergence of old scourges, such as malaria and tuberculosis, confirm the two contrasting faces of human infectious diseases at the end of the twentieth century.

The paradox raises two questions. First, how do epidemic diseases emerge and can we trace their geographical origins to any particular part of the world? Second, why do more diseases appear to be emerging now and how far does this crudescence relate to the unprecedented changes in the global environment?

To approach such questions implies a shift in gear compared to the pre-vious lecture. There, I concentrated on painting on a small, even a minia-ture, canvas: a single disease, only two small islands, a narrow time span.

Table 3.1. Examples of emerging virus diseases

Virus family	Virus	Signs/symptoms	Natural host	Geographical range
Orthomyxoviridae	Influenza	Respiratory	Fowl, pigs	Worldwide
Bunyaviridae	Hantaan, Seoul and other Hantaviruses	Haemorrhagic fever + renal sydrome + respiratory stress	Rodent (e.g. *Apodemus*)	Asia, Europe, US
	Rift Valley fever	Fever + haemorrhage	Mosquito, ungulates	Africa
Flaviviridae	Yellow fever	Fever, jaundice	Mosquito, monkey	Africa, South America
	Dengue	Fever + haemorrhage	Mosquito, human/monkey	Asia, Africa, Caribbean
Arenaviridae	Junin (Argentine HF)	Fever, haemorrhage	Rodent (*Calomys musculinus*)	South America
	Machupo (Bolivian HF)	Fever, haemorrhage	Rodent (*C. callosus*)	South America
	Lassa fever	Fever, haemorrhage	Rodent (*Mastomys natalensis*)	Africa
Filoviridae	Marburg, Ebola	Fever, haemorrhage	Unknown; ? Primate	Africa
Retroviridae	HIV	AIDS	? Primate	Worldwide
	HTLV	Adult T-cell leukaemia, neurological disease	Human virus (? originally primate virus)	Worldwide, with endemic foci

Note. HF, haemorrhagic fever; HIV, human immunodeficiency virus; HTLV, human T-cell leukaemia/lymphoma virus.
Sources: Haggett, 1994, *op. cit.* [note 1], Table 1, p. 93. Based in part on S. S. Morse, *The Evolutionary Biology of Viruses.* New York: Raven Press, 1994, Table 1, p. 326.

The brush I used in painting an epidemic was fine-tipped with even individual index cases identified. But in this third lecture I wish to set those constraints aside. The palette will now be expanded to many diseases, small islands will be replaced by global questions, narrow time bands will be expanded—backwards to the origins of human populations, forwards into this century. The fine-tipped brush will be put aside and replaced by a broad and coarse paint brush.

The geographical question of disease origins

I begin with the first question of disease origins. This can be approached in various ways and here I wish to concentrate on the geographical aspects. In what parts of the world did individual diseases emerge and what were the special conditions, if any, that may have contributed to that emergence?

Let me say at the outset that the region of origin of very few of the major human diseases is known with confidence. In a recent authoritative review of human diseases, the Australian epidemiologist Oliver Lancaster cites the cholera Vibrio which appears to have arisen in the Indian subcontinent (see Figure 3.1), bubonic plague and malaria in Africa or Asia, and yellow fever in Africa.[4] Claims that syphilis had a New World origin seem likely to be unfounded, and even diseases first recognized there (e.g. Tularemia, first described in Tulare county, California) have Old World analogues. The first hard evidence for most diseases comes many, many centuries after its probable emergence.

There has also been some research into the geographical origins of influenza with the Asian mainland, notably China but also southern Russia, implicated in some of the major epidemics of the last two centuries. Figure 3.2 shows the origins of the 1957–8 influenza A pandemic which had its origins in southern China. China has been implicated in a number of the genetic shifts that trigger global pandemics, although whether this is a probabilistic effect due to the high population densities there or intimate contact with suitable animal vectors is not known.[5]

By contrast with the old diseases, the new diseases offer more hope of a solution. The emergence of AIDS as a major human disease over the last decade has been accompanied by an unprecedented torrent of literature, some of it of geographical interest.[6] Notwithstanding, one still unsolved question is when and where did the epidemic of AIDS begin? (see Figure 3.3.) We know that the epidemic spread of the disease in the United States began in 1981, but there is increasing evidence that AIDS was present in a non-epidemic form for many years before the first epidemic cases. For example, Huminer and colleagues have identified a score of cases between 1953 and 1981 that meet the strict US Centers for Disease Control criteria for AIDS.[7] Other scholars have argued for a much older origin.

AIDS is, however, not the only 'new' disease epidemic of previously unrecognized diseases that have occurred during the last fifteen years. These include Lyme disease, legionnaires' disease, and toxic shock syndrome, all first recognized in the United States. Still others are the diseases

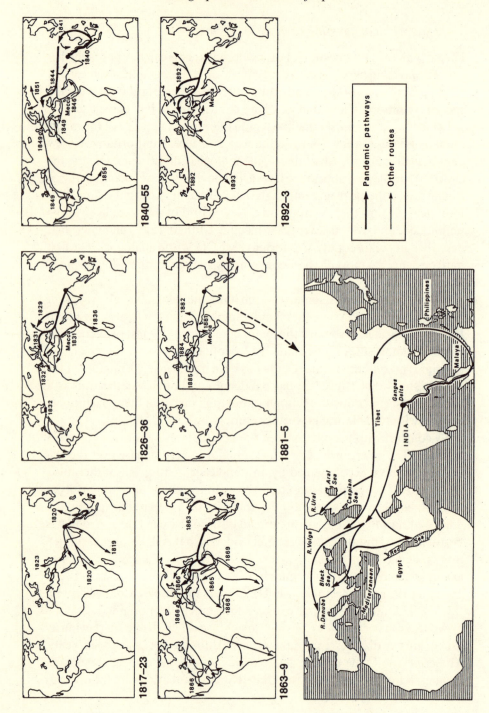

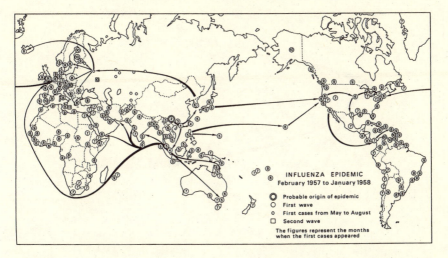

Fig. 3.2. Origin and spread of an influenza pandemic. Global sequence of spread of the 1957–8 pandemic is shown by vectors with figures indicating the month when the first cases appeared. The pandemic appears to have originated in southern China and spread by two pathways: westwards via the trans-Siberian corridor into eastern Europe and by sea from Hong Kong to Singapore and Japan.
Source: Cliff and Haggett, 1988, *op. cit.* [note 4], Fig. 6.4B, p. 237.

associated with the Lassa, Ebola, and Marburg viruses emerging from central Africa.

One indicator of the expanding number of diseases now recognized is provided by examining past medical literature. In 1917, the American Public Health Association published its pioneer handbook, *Control of Communicable Diseases in Man*. It listed control measures for thirty-eight communicable diseases, all those then officially reported in the United States. Since then, the number listed has steadily expanded so that the most recent edition, the fifteenth, now lists some 280 diseases.[8]

Both modern and historical examples of new infectious diseases share a number of features in common: (a) the onset of the new diseases appears to be sudden and unprecedented; (b) once the disease is recognized, isolated cases are retrospectively identified that occurred well before the outbreak;

Fig. 3.1. Origin and worldwide spread of cholera. World maps of six cholera pandemics in the nineteenth century showing the main pathways followed.
Source: Cliff and Haggett, 1988, *op. cit.* [note 4], Fig. 1.1D, p. 5.

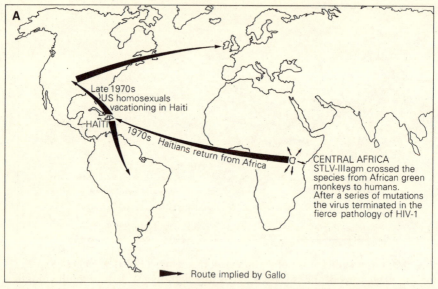

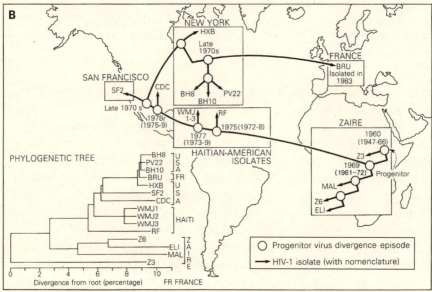

Fig. 3.3. The global diffusion of HIV-1. (*Above*) Shannon and Pyles model for the global diffusion of HIV-1. (*Below*) The initial spread of the HIV-1 virus derived by Li and co-workers from phylogenetic histories of virus isolates. This suggests that the earliest event is estimated to have occurred in Zaire, central Africa, around 1960 when the isolate Z3 diverged from an African progenitor virus. STLV-III agm refers to the simian T-cell lymphotrophic virus, a member of the retrovirus family.

Source: Smallman-Raynor *et al.* 1992, *op. cit.* [note 6], Figs 4.1B, 4.1C, pp. 145, 146.

and (c) previously unknown pathogens or toxins account for many of the new infectious diseases.

Ampel suggests four factors that may explain these observations:[9]

1. The infection was present all along but was previously unrecognized. As noted above, this is true to some extent for all four of the new diseases. However, it is unlikely that toxic shock syndrome, Lyme disease, or AIDS existed in their present forms in a large number of patients long before their recognition.

2. Pathogens responsible for these new diseases existed in the past but in a less virulent form. Some event, such as a genetic mutation, converted the organism to its virulent form. Rosqvist and co-workers showed that double-point mutations in the bacterium *Yersinia pseudotuberculosis* resulted in a marked increase in the virulence of this organism *in vitro*.[10] Carmichael and Silverstein postulated that the marked increase in mortality associated with smallpox in sixteenth-century Europe could have been due to mutations in the causative virus, variola.[11] Such events could also have occurred with regard to HIV-1. Smith and colleagues examined the genomic sequences of conserved areas of HIV-1 and compared these to human immunodeficiency virus type 2 (HIV-2), the related and potentially less virulent strain, and to the simian immunodeficiency virus, which is virulent for non-human primates but not for humans. They conclude from these data that 'HIV-1 diverged from HIV-2 in 1951, plus or minus three years, and that the onset of conspicuous diversification of these viruses correlates strongly with the known historical rise of the pandemic . . .'.[12]

3. A new epidemic arises from the introduction of a virulent organism into a non-immune population. The epidemic of smallpox that decimated the Aztec population of Mexico in 1520 is one historically important example. We noted in the previous lecture the first introduction of measles into Fiji in 1875 in which one-quarter of the islands' populations died within a two-month period.[13]

4. Environmental and behavioural changes provide a new environment in which the disease-causing organisms may flourish. Legionnaires' disease is related to the growth of cooling towers and evaporative condensers in the 1960s; Lyme disease to the growth of the deer population in the woodlots that have grown in abandoned fields of New England. Toxic shock disease is related to a behavioural change, the

increasing use of tampons by menstruating women. Further back in time, the types of changes that accompanied agriculture may well have brought shifts in disease patterns. One hypothesis needing study is that malaria began to attack humans 10,000 years ago when Africans shifted from hunting on the savannah to farming in the forests.

To set the origin in context we should note the many cases where diseases probably failed to emerge from contact with a disease-bearing organism. In the case of viruses, McKeown has noted that when an infection comes into contact with a strong human host three events may occur: (1) the virus may fail to multiply and the encounter passes unnoticed; (2) the virus multiplies rapidly and kills the host without being transmitted to another host; or (3) virus and host populations (after a period of adaptation) settle down into a prolonged relationship which we associate with sustainable diseases.[14] The relative proportion we can attach to the three outcomes is unknown but there is good reason to think the third is the rarest.

The limits of historical evidence

So far, we have been discussing those diseases that have emerged in recent decades. But as we move backwards through time we continue to find diseases that have plagued the human population for all of recorded history. For these diseases two lines of reconstruction are possible.

The first is historical. In the second lecture, we limited ourselves to the epidemiological records of two islands which allowed us to go back in time for a little over a century. Because of problems of accurate identification and recording conventions it is difficult to press back such 'hard' quantitative evidence much before the 1880s. There are a few exceptions, notably in the historical statistics gathered for some our large cities and some unique Japanese records, as Figure 3.4 demonstrates.[15]

But if quantitative evidence is scarce there is a widening range of other historical and pre-historical sources. The early flowering of palaeopathology as a science came largely from the Egyptologists. Study of more than 8,000 mummified bodies revealed traces of various infectious diseases: body scars resulting from smallpox, skin infections, and specific diseases, such as tuberculosis and leprosy, were clearly recognizable. Traces of parasitic disease-causing organisms have been found in faeces and middens, both in Egypt and in other areas where mummified bodies have been exam-

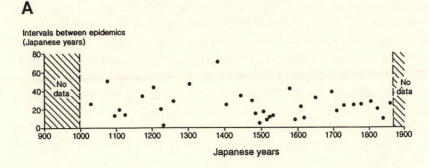

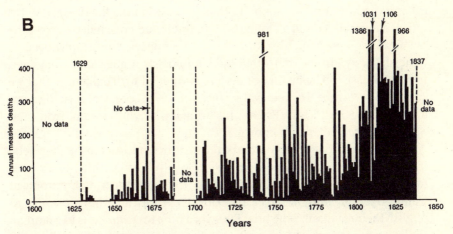

Fig. 3.4. Historical records of epidemics. (*Above*) Japanese records of measles epidemics from 1000 to 1840. Distribution of epidemic dates plotted against the time interval between epidemics. (*Below*) Annual measles deaths in London, 1600–1850.

Source: Redrawn from tables in Janetta, 1987, *op. cit.* [note 15], Table 5.1, p.117; Creighton, 1965, *op. cit.* [note 15], pp. 634–60; *passim* in Cliff *et al.* 1993, *op. cit.* [note 15], Figs 3.2, 3.4, pp. 53, 57.

ined (e.g. Peru, 3000 BP).[16] Bodies preserved in bogland and in icefields are smaller in number but are also yielding disease-related information.

Where only skeletal remains are found in archaeological sites interpretation causes greater problems as a series of microorganisms can produce very similar changes. But osteomyelitis, tuberculosis, leprosy, and syphilis have been identified.

Once we enter the period of historical evidence, a wider range of diseases can be recognized. For example, a bas-relief dating from the eighteenth

Egyptian dynasty (about 3500 BP) shows the shrivelled leg of a priest, characteristic of a patient recovered from paralytic poliomyelitis.[17] Medical treatises from the Greek, Roman, Indian, Chinese, and Arab civilizations describe diseases which we try to pair with their modern counterparts. I say try, because although past diseases may be recorded in historical accounts it is not always clear which disease is in fact being described. The first recorded outbreaks of a new epidemic disease occurred in Athens, Greece, from 430 BC to 427 BC. Thucydides' account has been much studied by epidemiologists: he describes an epidemic which (a) occurred in two distinct waves, (b) had a devastating impact with a mortality of around one-third, (c) appeared to cause immunity in those affected in the first wave, and (d) affected animals as well as humans. Langmuir and colleagues suggest the outbreak was probably due to influenza complicated by staphyloccocal toxic shock syndrome, but other epidemiologists argue that it may be a unique disease.[18] The history of disease is studded with epidemics that we can no longer trace. In medieval France, we have records of a vicious epidemic termed the 'Picardy Sweat'. Just what it was, where it came from, and why it disappeared are not known.

A second approach is to outflank the historical records through genetic tracing. The revolution in molecular biology of the last fifty years owes much to two young men working in the Cavendish Laboratory at Cambridge, England, in the early 1950s. By April 1953 they were able to publish their results and the brief 900-word note to *Nature* was to set off an explosion of research which was to transform our understanding of biology and the origins and evolution of life itself.[19]

Ripples from that explosion are now reaching towards the kinds of global questions that we raise here. Studies of DNA of the human cell gives us clues about our evolutionary past that take us back to the beginnings of humanity 100,000 years ago and to the origins of life itself 3,000,000,000 years before that.[20] Each cell of the human body contains about 2 metres of DNA, each containing some 3 billion coded letters which makes every one of us unique, different in measurable ways from our fellows. In a real sense, molecular biology is starting to confirm the fact that, despite superficial likenesses, each of us is a unique living fossil.

In the long association between humankind and disease, various protective devices have been evolved and inherited. Malaria is caused by a single-celled parasite that is transplanted into humans by mosquitoes. The malaria parasite does much of its damage by invading red blood cells but in West Africa, many people carry a gene that alters the red cell so that the parasite cannot attach to it. Relevant cell surface antigens occur in up to half the population of West Africa but are rare elsewhere in the world. There are, however, isolated populations elsewhere that have some protective tricks. One of the most intriguing is the population of Coruche, a town in central Portugal, where malaria was once common. Although the people there have most of their DNA in common with other Portuguese, the DNA around the sickle-cell gene is of a type found only in West Africa. Jones suggests that children from inter-racial unions were brought back from Africa to Portugal.[21] Over the last three centuries the irrelevant skin-colour genes were lost but the gene giving protection against the local threat of malaria was retained and spread.

Coruche stands as a marker for the detailed genetic geography which can now be built up of the human population spread around the Earth. Not only does genetic research at the molecular level allow the sorting out of whole populations, it also gives clues as to the diseases that our ancestors may have suffered over the 6,000 generations since humankind first appeared on earth.

Carl Sauer and geographical speculation

A third line of approach to probing into disease origins in which the evidence is obscured is to transpose our present understanding of disease processes into past environments. Even if we concede that for most of the existing epidemic diseases of humans, the question of origins lies too far in the past to be resolved the question need not end there. The barrier to tracing origins is not however a new one. The Berkeley geographer, Carl Ortwin Sauer, addressed a similar problem of inaccessibility in tracing the origins and spread of agriculture, a question addressed in his Bowman Lectures to the American Geographical Society (published as *Agricultural Origins and Dispersals* in 1952).[22] Sauer, already a medallist of the Society, was invited to give the second Bowman series and specifically to link it to another of Bowman's research interests, pioneer settlements. Sauer accepted the invitation but, typically, interpreted 'pioneer man' to span an earlier and wider range of time than perhaps either Bowman or the Trustees might have had in mind. His aim was nothing less than a global survey of arguably the greatest pioneering revolution in human history, the beginnings of agriculture.

But Sauer was not interested in providing what he called dismissively a 'well-polished abstract of accepted learning'. Then at the age of sixty-two and only five years from retirement from Berkeley, Sauer saw this as an opportunity to speculate, to 'set out a prospectus of knowledge which is not clearly within our grasp' and to identify those frontier areas on which he saw 'good prospects of new learning'.[23]

Although Sauer himself was dismissive of methodology and regarded it as a 'habit-forming drug' his Bowman Lectures were in fact strongly methodological.[24] There are two aspects to this methodology that stand out: the use of deductive locational principles and the use of maps as hypotheses.

Sauer called his principles 'certain basic premises'. There were six of them:

1. 'Agriculture did not originate from a growing or chronic shortage of food.' The improvement of plants by selection for better utility to man was accomplished only by a people who lived at a comfortable margin above the level of want.

2. 'The hearths of domestication are to be sought in areas of marked diversity of plants or animals, where there were varied and good raw materials to experiment with, or in other words, where there was a large reservoir of genes to be sorted out and recombined.' This led him to the view that agriculture would begin in well-diversified terrain that perhaps also had some variety of climate.

3. 'Primitive cultivators could not establish themselves in large river valleys subject to lengthy floods and requiring protective dams, drainage, or irrigation.' Sauer was concerned by the potamic thesis of agriculture originating in the great valleys of the Near East and favoured Vavilov's view that hill and mountain lands were more likely hearths.

4. 'Agriculture began in wooded lands.' Sauer objected to an earlier European view that the loess lands were centres of neolithic agriculture because they were grassy. He argued that grasses were much more difficult to cultivate than woodlands.

5. 'The inventors of agriculture had previously acquired special skills in other directions that predisposed them to agricultural experiments.' Sauer doubted if the transition from mobile hunting groups to plant breeding were feasible.

6. 'Above all, the founders of agriculture were sedentary folk.' Growing crops require constant attention, and Sauer recalled from his own field experience that he had never seen primitive plantings that were not closely watched over until the crop is secured.

The maps in the monograph are the core of Sauer's results. I shall refer only to the second (see Figure 3.5). This shows the cores of agriculture in the New World: an earlier tropical hearth for planting based on the northern Andean valleys, and a later hearth for seed planters based on southern Mexico. Again, the lines of diffusion north up into the Mississippi Valley and south and east along the Andean chain and into the Amazon basin are plotted as vectors spreading out from the hearths.

Each of Sauer's six ecological principles led to either a spatial inclusion or a spatial exclusion of possible hearths for agriculture. The results of the sieving process resulted in a series of maps, some using highly original projections to give an effective global picture. Sauer saw the maps as

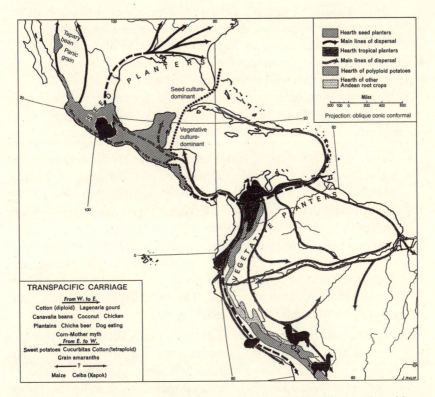

Fig. 3.5. Sauer's suggested origins of agriculture in the New World. The evidence leading Sauer to this suggested hearth area in the American tropics is discussed in the text.

Source: Sauer, 1952, *op. cit.* [note 22], Fig. 2, p. 42. (Copyright: American Geographical Society, 1952.)

hypotheses and commented in a letter to Gladys Wrigley that '. . . they are educated guesses . . . we cannot identify with assurance the place of origin of any domestications'.[25] Sauer stresses that the maps are intended to set forth what was, in terms of present knowledge, reasonable interpretation; they were as he put it '. . . worksheets to be revised as better knowledge comes to hand'.[26]

Extension of Sauer's methods to disease origins

Carl Sauer's interest was in the origins and dispersal of agriculture; my own interest is in a different and apparently distant field. But some of the methods he uses are of interest. Perhaps the most thoughtful part of the essays comes where he eschews this evidence and argues deductively from a series of principles. In doing so, Sauer was following on a long tradition within archaeology. A century before, Francis Galton had set out the six conditions under which wild animals may become domesticated.[27]

Although it is premature to follow Sauer's methodology and search for some locational principles that may help identify candidate areas for the origin of epidemic diseases, some provisional points can be made:

1. Since many epidemic diseases of humans are what Greenwood termed 'crowd diseases', they depend on the clustering of the host population into densities able to sustain an infection chain.[28] We noted in the first lecture that the precise threshold population will depend on a number of factors, notably the latent interval of the disease and its infectiousness. The longer a disease has a hold on someone, and the more efficiently it is transmitted, the smaller the population needed to allow it to persist. Thus, Bartlett showed for measles that with a latent interval of fourteen days it needs a community of at least 300,000 people to keep going.[29]

2. We have lived in groups as large as this for only about 2,000 years, so measles must be a fairly new disease. The implications of (1) are that most infectious diseases of humans first occur late in human history when agriculture allowed sedentary populations at sufficient threshold densities to occur.

3. The crossover of infections from animal populations (zoonoses) suggests that the close contact coming from domestication of animals is an important milestone. There are today some molecular similarities

between the measles virus and the virus responsible for both distemper in dogs and rinderpest in cattle. Fenner and Fiennes have discussed other animals from which human infectious diseases have come: many rhinoviruses from horses which cause the common cold; respiratory tuberculoses from the bovine type present in wild cattle; leprosy from the water buffalo; diphtheria from cattle; syphilis from monkeys.[30] Mumps and smallpox are also believed to be crossovers from animal populations but the original host animals are not known.

4. Crossover of infections from other animal species (both domesticated and non-domesticated) points to the greater species diversity of organisms of all kinds in the tropics, particularly the humid regions. Viruses, the most diverse of all 'life' forms, and bacteria reflect this strong skew towards the tropics.

5. The critical distinction between Old and New World plant and animal species which lies at the heart of Sauer's thesis, has its counterpart in epidemic diseases. Cook and Lovell have traced the transatlantic introduction of measles from Spain into New Spain.[31] They reported a study of disease outbreaks from Seville, suggesting that measles was virtually endemic in the port city, from which most official trade with the Indies was conducted. By 1531, measles first appeared in central Mexico in pandemic form, reappearing thereafter in cycles of approximately thirty years. Similar reports trace the introduction of measles throughout New Spain: Guatemala (1532), Nicaragua (1533), Peru (1535). By far the most devastating outbreaks of sickness, however, swept the Andes from 1585 to 1591, when several imported diseases, measles among them, were present.

6. Islands are unlikely sources of disease origin. Their populations were usually too small and their locations too remote for any new diseases to be either sustained or spread; the historical evidence points to epidemics consistently being introduced from outside. This does not preclude some rare diseases being sustained on some larger islands (kuru on New Guinea). A comprehensive review is given in *Island Epidemics*.[32]

7. Improved transport has a critical impact on disease diffusion. This is an area in which geographers have made significant contributions. Pyle's study of the way in which the evolving transport links in eastern North America affected the differential spread of cholera in the

epidemics that started in 1832, 1848, and 1866 is a classic example.[33] For Fiji, we noted in the previous lecture how the switch from slow-moving small sailing vessels to larger and faster steamships directly affected the introduction of the measles virus.[34] Sailing vessels bringing migrant workers dominated the India–Fiji route from 1879 until the beginning of the twentieth century, and the voyage length of fifty to ninety days demanded too long a measles infection chain (around five or six virus generations) to allow the virus to reach the destination. Steamships slashed the voyage time to around thirty days (about two virus generations) with the effect that half the vessels which left Calcutta or Madras with measles on board still had infected cases when Suva was reached. In the last thirty years, the vast increase in the speed and volume of air passenger traffic has had an equally critical impact on intercontinental virus movements.

One implication of these principles is that some of our classic epidemic diseases may themselves have originated at the same time and in the same location as Sauer's agricultural hearths. Disease agents were in some senses 'camp followers' that joined, unbidden, in the domestication process.

We have used Sauer's methods to try and establish a probable area of origin for measles.[35] This is based on the known evolution of high density population areas in the Old World coupled with Bartlett's theoretical endemicity hearth. We argue for measles to become established in the great river valley civilizations of the Tigris and Euphrates by around 3000 BC with domesticated animals (notably the dog and cattle) as probable sources (see Figure 3.6).

Global change and its disease implications

Although the investigation of disease origins and dispersal has an innate fascination their study is not entirely one of just antiquarian interest. Disease patterns are changing at an unprecedented rate today and raise our second question. Why do more diseases appear to be emerging now than in the past and how far is this crudescence relate to the unprecedented changes in the global environment?

Diseases spread within a specific historical and geographical context. At the beginning of the twenty-first century, the environmental context within which disease control is set is changing at a faster rate than any time in

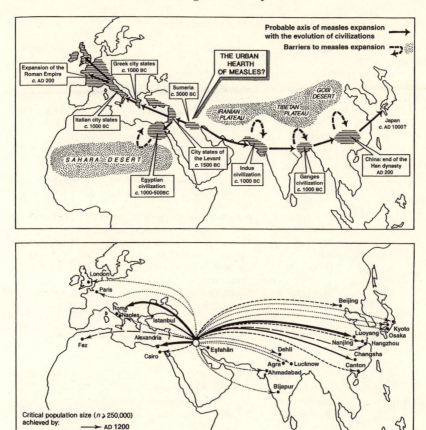

Fig. 3.6. Hypothetical reconstruction of original hearth of measles and its spread. (*Above*) Probable axis of measles diffusion from its Middle East hearth, and physical barriers to measles expansion. (*Below*) Evolution of settlements at, or above, the Bartlett threshold (approximately 250,000 people) for measles endemicity. Vectors link settlements with a probable measles hearth in Sumeria.
Source: Cliff *et al.* 1993, *op. cit.* [note 15], Fig. 3.1, p. 51.

human history. Table 3.2 summarizes some of those environmental changes that have disease implications.

We illustrate here the impact of four such changes: (1) demographic growth of the host population, (2) the collapse of geographical space, (3) global land-use changes, and (4) global warming, and their joint interactions.

Table 3.2. Geographical changes and virus emergence

Geographical change	Disease	Probable mechanism	Location
Increased spatial interaction	Dengue	Disseminated by travel and migration.	Worldwide
	Yellow fever	Both virus and major vector (*Aedes aegypti*).	Africa, Caribbean
	Seoul-like viruses	Infected rats carried on ships.	US
Land-use change: (A) Agricultural	Influenza	Integrated pig–duck farming.	China
	Hantaan	Contact with rodents during rice harvest.	China
	Argentine HF	Agriculture favours natural rodent host; human contact during harvest.	South America
	Bolivian HF	Contact with rodent host during harvest.	South America
	Oropouche	Cacao hulls encourage breeding of insect vector.	South America
	Monkeypox	Subsistence agriculture and hunting in forests; increased contact with rodent host.	Tropical Africa
(B) Forests and woodland	Kyasanur forest disease	Tick vector increased as forest land replaced by sheep grazing.	India
	Lyme disease	Tick vectors increased as fields replaced by woodland.	North-east US
(C) Water	Dengue, dengue haemorrhagic fever, Yellow fever	Water containers encourage breeding of mosquito vector.	
	Venezuelan equine fever, encephalitis, Rift Valley fever	Building of dams and irrigation favour increase in vector.	Panama, Africa
Military operations	Leishmaniasis	Military action in new environment (e.g. operation 'Desert Storm').	Persian Gulf
Global warming	Malaria	Extension of thermal range into middle latitudes.	Worldwide

Note. HF, haemorrhagic fever.
Source: Haggett, 1994, *op. cit.* [note 1], table 2, p. 97. Based in part on S. S. Morse, *The Evolutionary Biology of Viruses*. New York: Raven Press, 1994. table 3, p. 330.

Growth and relocation of the human population

Whatever the rate of past disease emergence, there are reasons to consider the early twenty-first century as one of special significance for the human host population. Figure 3.7 shows the historical pattern of growth in the human population from 1750 with a forward projection to 2100.

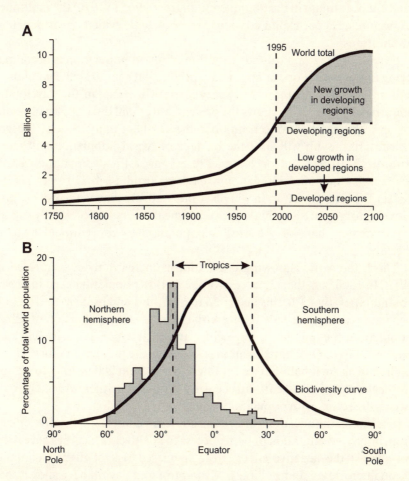

Fig. 3.7. Global population growth. (A) Global population growth. Course of global population growth for the period from 1750 and projected forward to 2100. (B) Geographical distribution of population distribution in terms of 5° latitudinal bands north and south of the equator. The biodiversity curve is approximate and does not make allowances for the global distribution of humidity.
Source: Haggett, 1995, *op. cit.* [note 1], Fig. 3, p. 101.

Three aspects of this change call for comment. First, the rapid acceleration in growth is very recent. In the past four decades, the world's population has more than doubled, from 2.5 billion in 1950 to 5 billion in 1988. On the United Nations 'medium-growth' assumptions, this total is expected to reach 6.3 billion by the beginning of this century and 8.5 billion by the year 2025.[36] Although that rate of growth is now decelerating (its peak was at 2.1 per cent per annum in the quinquennium from 1965 to 1970) the multiplier of resource use per capita continues to grow with evident environmental implications.

Second, a geographical redistribution of world population is accompanying this growth. For example, it is expected that some 94 per cent of population growth over the next twenty years will occur in the developing countries. Figure 3.7(B) shows the present latitudinal distribution of population with a marked concentration in the northern mid-latitudes. Present and future growth will shift the balance of world population still more towards the tropics and low latitudes (e.g. I have estimated that the average temperature of global population will rise by around +1 °C, from 17 °C to 18 °C).[37] This concentration will place more people than in the world's previous history in areas of high microbiological diversity, potentially exposing a greater share of the world's population to conventional tropical diseases.

Third, the world's growing population is increasingly concentrated in cities. In 1800, less than 2 per cent of the world's population lived in urban communities. By 1970, this had risen to one-third and by 2000 the fraction will have reached one-half. Along with the increasing proportion of urban population, the number of large cities and their average density will also have increased. On United Nations estimates, the number of cities with a million or more inhabitants is expected to rise from 200 in 1985 to 425 by the beginning of the twenty-first century. At that date there are likely to be twenty-five cities with populations in excess of 11 million.

The disease implications of urbanization are complex.[38] Positive effects from improved sanitation or better access to health care facilities have to be set against the negative effects from increased risk of disease contacts through crowding and pollution. Crompton and Savioli have shown that where rural–urban migration in developing countries results in peri-urban shanty settlements, high rates of intestinal parasitic infections (notably amoebiasis, giardiasis, ascariasis, and trichuriasis) can result.[39] Each are common intestinal infections caused by protozoan parasites or helminths transferred from human to human by the faecal–oral route. They pose an increasing health burden as the share of urban populations in developing countries rises towards one-half of the total population. In the long-term historical context, Haggett has shown that the aggregation of human populations into high density urban 'islands' had important effects in providing the host reservoirs for maintaining infection chains.[40] We have discussed the implications for urbanization for the spread of measles elsewhere, as has Fine for a wide range of other infectious diseases.[41] But in no disease has the metropolitan structure played a more important role in disease spread than AIDS: Gould's maps (reproduced in Figure 3.8) show the strong effect of the urban hierarchy on the rapid spread of AIDS in the United States over the period 1984–90.[42]

Changing global land use

The combination of population growth and huge technological changes has given mankind capacity to alter environments in ways that are unprecedented in human history. We illustrate the disease implications of three such changes.

Accelerated world population growth has put pressure on food supplies in tropical areas and has led to the colonization of new environments in the search for expanded food production. Venezuelan haemorrhagic fever is a severe and often fatal zoonotic virus disease only recently identified in the Guanarito area in central Venezuela. Cases were not found in the cities but confined to rural inhabitants of the area who were largely engaged in farming or cattle ranching. Major outbreaks in 1989 and again in 1990–1 had fatality rates of around one-quarter. First diagnosed as due to dengue haemorrhagic fever, the disease is now known to be due to a separate virus, named the Guaranito virus, which is associated with rodent reservoirs.

Guaranito appears to be one of a family of arenaviruses known to cause haemorrhagic fevers in humans. They include the Junin and Machupo viruses associated with haemorrhagic fever outbreaks in Argentina and Bolivia. In each case transfer appears to be from a wild rodent host (*Akodon azarae* and *Calomys musculinus* in Argentina and *C. callosus* in Bolivia) with the main risk associated with exposure during the corn harvesting season. Similar seasonal risks from epidemics of haemorrhagic fevers are associated with the family of Hantaan viruses in China which appear to be transferred to humans during the rice harvest. Field mice, rats, and bank voles are involved in fever transmission in different parts of the world.

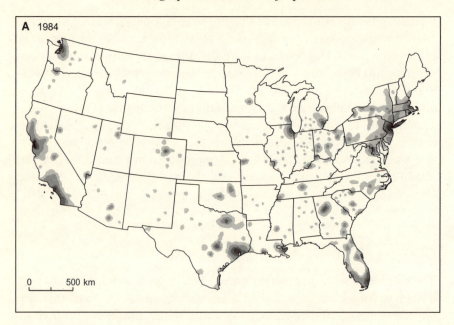

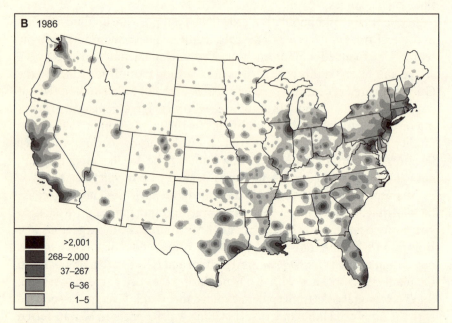

Fig. 3.8. Transmission of AIDS through the United States urban hierarchy. The changing intensity of the AIDS epidemic in 1984, 1986, 1988, and 1990. The effects of hierarchic diffusion are already evident by the first map (A), in the second (B), the disease has moved down to smaller centres, in the third (C), spatially contagious diffusion becomes more prominent as urban commuter fields pump the virus

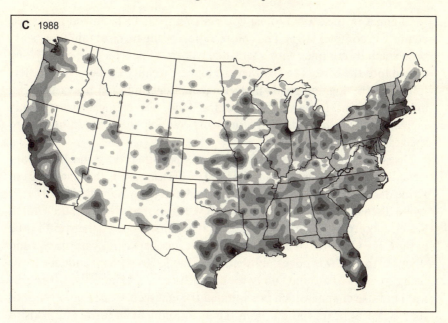

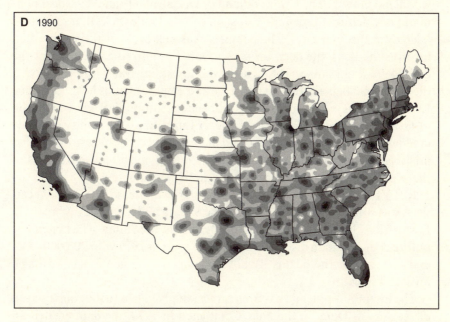

into the suburbs of regional epicentres. In the last map (D), clear alignments along interstate highway systems are intensifying, and earlier urban 'beads on a string' are reaching out to coalesce.

Source: Gould, 1999, *op. cit.* [note 42], Plates 5–8.

Changes in the global forest cover also appear to be linked to disease changes in complex ways. The deforestation of the tropical rain forests has been spatially complex with a fern-like pattern of new logging roads being driven into the forests to abstract the highest quality timber. New settlers following the logging roads into Amazonia encountered heavy malarial infections. This is partly because the land use changes have greatly increased the forest-edge environments suitable for certain mosquito species.

Disease changes can also result from an opposite process in which abandoned farmland reverts to woodland. The classic case is the emergence of Lyme disease, caused by the spirochaetal bacterium *Borrelia burgorferi*.[43] Lyme disease is now the most common vector-borne disease in the United States but retrospective studies suggest it was not reported there until 1962 in the Cape Cod area of New England. The link between Lyme disease and 'Lyme arthritis' was not established until the 1970s with an endemic focus being recognized around Old Lyme in south-central Connecticut. The critical land-use change which precipitated the emergence or re-emergence of the disease appears to have been the abandonment of farmland fields to woodland growth. The new woodland proved an ideal habitat for deer populations which are the definitive host for certain Ixodes ticks that spread the bacterium through bites. The complex seasonal cycle of the vectors that involves the ticks, the deer, the white-footed mouse (the reservoir for the pathogen), and human visitors using the forest illustrates how sensitive is the ecological balance in which disease and environment is held. Epidemic Lyme disease is now an increasing problem in Europe, fuelled by reversion of farmland to woodland (partly due to European Community set-aside land policies), deer proliferation, and increased recreational use of forested areas. Lyme disease has now been reported from most temperate parts of the world in both the northern and southern hemispheres.

Rift Valley fever was until recently primarily a disease of sheep and cattle. It was confined to Africa south of the Sahara with periodic outbreaks in East Africa, South Africa and, in the mid 1970s, the Sudan. The first major outbreak as a human disease occurred in Egypt in 1977 with 200,000 cases and 600 deaths, the deaths usually associated with acute haemorrhagic fever and hepatitis.

The Egyptian epidemic has been provisionally linked to the construction of the Aswan Dam on the Nile. Completed in 1970, the dam created an 800,000-hectare water body and stabilized water tables so that surface water provided breeding sites for mosquitoes. Whether the mosquito population provided a corridor for allowing the virus to enter Egypt from the southern Sudan has yet to be proved. But the possibility leads to concern for the epidemiological implications of other dam-building schemes in the African tropics. Completion of the Diama Dam on the Senegal River in 1987 was followed by a severe outbreak of Rift Valley fever upstream of the new dam. Over 1,200 cases and 244 deaths resulted. But in contrast to Egypt, immunological studies showed Rift Valley fever to be already endemic in people and livestock in a wide area of the Senegal River basin. Ecological changes favouring the vector and associated with dam-building seem to be implicated in allowing both: (a) invasion of the virus into a previously virgin population, and (b) severe flare-ups in a population with low level endemicity.

Global warming

Of the many global scenarios for disease and the environment in the early part of this century, it is the health implications of global warming that have caught the attention of governments and press worldwide. There have already been major studies of its potential health implications in at least three countries: the United States, Australia, and the United Kingdom. The World Health Organization also has a committee looking at this issue.[44]

A number of health effects have been postulated as following from a worldwide increase in average temperature from global warming. For infectious diseases, the main effects relate to changes in the geographical range of pathogens, vectors, and reservoirs. So far, few attempts have yet been made to compute the relative burden of morbidity and mortality that would result from these effects. Any such calculation would also need to offset losses against gains that might accrue (e.g. reduction in hypothermia against increase in hyperthermia).

The magnitude and spatial manifestations of global warming are still speculative. One of the main conclusions of the report of the Intergovernmental Panel on Climate Change (IPCC) in 1990 was how far research still had to go before reliable estimates of global warming could be identified. But some rough orders of magnitude can be computed from the estimates of the different models that have been used. In global terms, warming appears to range from '. . . a predicted rise from 1990 to the year 2030 of 0.7 °C to 1.5 °C with a best estimate of 1.1 °C'.[45]

We can obtain some idea of the implications of the predicted shift for local mean temperatures with reference to the United Kingdom. Current differences between the coldest (Aberdeen, latitude 57.10 N) and warmest (Portsmouth, latitude 50.48 N) of its major cities is 2.4 °C; this is well beyond the postulated IPCC warming effect by the year 2030. Climate is a much more complex matter than average temperature, but—if the global warming models carry over to the United Kingdom—then, by 2030, Edinburgh might have temperatures something like those of the English Midlands, and London something like those of the Loire Valley in central France. If we accept the much higher estimate of +4.8 °C warming over eighty years, this brings London into the temperature bands of southern France and northern Spain. Provided that these projections are sensible, something might be gained by comparative studies of disease incidence within the United Kingdom and adjacent European countries, and disease incidence in warmer climates that match those predicted for the United Kingdom.

The biological diversity of viruses and bacteria is partly temperature-dependent, and it is much greater in lower than higher latitudes (see Figure 3.7). Conditions of higher temperature would favour the expansion of malarious areas, not just for the more adaptable *Plasmodium vivax* but also for *P. falciparum*. Rising temperatures might also allow the expansion of the endemic areas of other diseases of human importance: these include, for example, leishmaniasis and arboviral infections, such as dengue and yellow fever. Higher temperatures also favour the rapid replication of food-poisoning organisms. Warmer climates might also encourage the number of people going barefoot in poorer countries, thereby increasing exposure to hookworm, Schistosoma, and Guinea worm infections. But not all effects would be negative. Warmer external air temperatures might reduce the degree of indoor crowding and lower the transmission of influenza, pneumonias. and 'winter' colds.

While modest rises in average temperatures are the central and most probable of any greenhouse effects, they are likely to be accompanied by three other main changes: (1) sea level rises of up to a metre, (2) increased seasonality in rainfall, thus reducing the level of water available for summer use, and (3) storm frequency increases.[46]

The collapse of geographical space

The second main environmental change has come from the collapse (in terms of both time and cost) of geographical space and the increased spatial mobility in the human population that has accompanied such a collapse. We look first at the evidence for such change and then at its disease implications.

Ways in which travel patterns have changed for the host population over recent generations have been shown in an interesting way by the distinguished epidemiologist, David Bradley of the London School of Hygiene and Tropical Medicine.[47] Bradley compares the travel patterns of his great-grandfather, his grandfather, his father, and himself (see Figure 3.9). The life-time travel track of his great-grandfather around a village in Northamptonshire could be contained within a square of only 40 km per side. His grandfather's map was still limited to southern England, but it now ranged as far as London and could easily be contained within a square

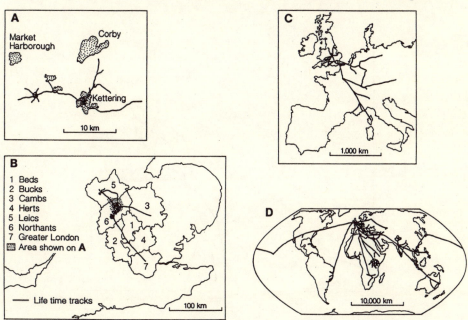

Fig. 3.9. Increasing travel over four male generations of the same family. (A) Great-grandfather. (B) Grandfather. (C) Father. (D) Son. Each map shows in a simplified manner the life-time tracks in a widening spatial context, with the linear scale increasing by a factor of 10 between each generation.

Source: Bradley, 1988, *op. cit.* [note 47], Figs 1–4, pp. 2–3.

of 400 km per side. If we compare these maps with those of Bradley's father (who travelled widely in Europe) and Bradley's own sphere of travel, which is worldwide, then the enclosing square has to be widened to sides of 4,000 km and 40,000 km, respectively. In broad terms, the spatial range of travel has increased tenfold in each generation so that Bradley's own range is one thousand times wider than that of his great-grandfather.

Against this individual cameo, we can set some broader statistical trends from recent years. One indicator of the dramatic increase in spatial mobility is shown in Figure 3.10. This plots for France over a 200-year period the average kilometres travelled daily both by transport mode and by all modes. Since the vertical scale is logarithmic the graph shows that, despite changes in the mode used, the average travel has increased exponentially broken only by the two World Wars. Over the whole period, mobility has increased by more than 1,000.

The precise rates of flux or travel of population both within and between countries are difficult to obtain in official statistics. But most available evi-

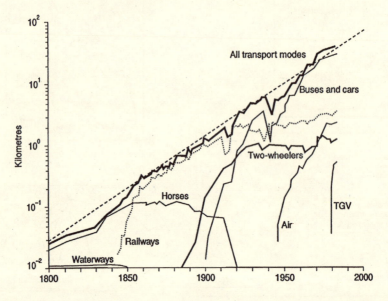

Fig. 3.10. Increased spatial mobility of the population of France over a 200-year period, 1800–2000. Note that the vertical scale is logarithmic so that increases in average travel distance increases exponentially over time. (TGV, Train à Grande Vitesse.)

Source: Haggett, 1994, *op. cit.* [note 1], Fig. 4, p. 102.

dence suggests that the flux over the last few decades has increased at an accelerating rate. While world population growth rate since the middle of the twentieth century has been between 1.5 and 2.5 per cent per annum, the growth in international movements of passengers across national boundaries has been between 7.5 and 10 per cent per annum. One striking example is provided by Australia: over the last four decades its resident population has doubled, while the movement of people across its international boundaries (i.e. into and out of Australia) has increased nearly one hundredfold.

Disease implications of global change

The implications of increased travel are twofold: short-term and long-term. First, an immediate and important effect is the exposure of the travelling public to a range of diseases not encountered in their home country. The relative risks encountered in tropical areas by travellers coming from Western countries (data mainly from North America and Western Europe) have been estimated by Steffen and Lobel and are given in Figure 3.11.[48]

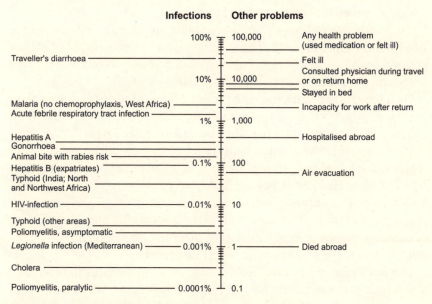

Fig. 3.11. Relative threats posed by communicable diseases to travellers in tropical areas. Note that the scale is logarithmic.
Source: Steffen and Lobel, 1995, *op. cit.* [note 48], Fig. 1, p. 56.

These suggest a spectrum of risks from unspecified 'travellers' diarrhoea' (a high risk of 20 per cent) to paralytic poliomyelitis (a very low risk of less than 0.001 per cent). Another way in which international aircraft from the tropics can cause the spread of disease to a non-indigenous area is seen in the occasional outbreaks of tropical diseases around mid-latitude airports. Typical are the malaria cases that appeared within 2 km of a Swiss airport, Geneva–Cointrin, in the summer of 1989.[49] Cases occurred in late summer when high temperatures allowed the in-flight survival of infected Anopheles mosquitoes that had been inadvertently introduced into the aircraft while at an airport in a malarious area. The infected mosquitoes escaped when the aircraft landed at Geneva to cause malaria cases among several local residents, none of whom had visited a malarious country.

A second short-term factor with modern aircraft is their increasing size. Bradley postulates a hypothetical situation in which the chance of one person in the travelling population having a given communicable disease in the infectious stage is 1 in 10,000.[50] With a 200-seat aircraft, the probability of having an infected passenger on board (x) is 0.02 and the number of potential contacts (y) is 199. If we assume homogenous mixing, this gives a combined risk factor (xy) of 3.98. If we double the aircraft size to 400 passengers, then the corresponding figures are $x = 0.04$, $y = 399$, and $xy = 15.96$. In other words, *ceteris paribus* doubling the aircraft size increases the risk from the flight fourfold. Thus, the new generation of wide-bodied jets presents fresh possibilities for disease spread, not only through their flying range and their speed, but also from their size.

On a longer time scale, increased travel brings some possible long-term genetic effects. With more travel and longer-range migration, there is an enhanced probability of partnerships being formed and reproduction arising from unions between individuals from formerly distant populations. As Khlat and Khoury have shown, this can bring advantages from the viewpoint of some diseases.[51] For example, the probability of occurrence of multifactorial conditions, such as cystic fibrosis or spinal muscular atrophy, is reduced; the risk of these conditions is somewhat higher in children of consanguineous unions. Conversely, inherited disorders such as sickle-cell anaemia might become more widely dispersed.

The last illustration in this lecture (Figure 3.12) captures the theme of global change in a summary form.[52] It shows the way in which the export of green monkeys from the African tropics (Uganda) into mid-latitude countries (research laboratories in Europe) brought a severe but limited outbreak of a rare haemorrhagic fever. The naming of the disease, Marburg

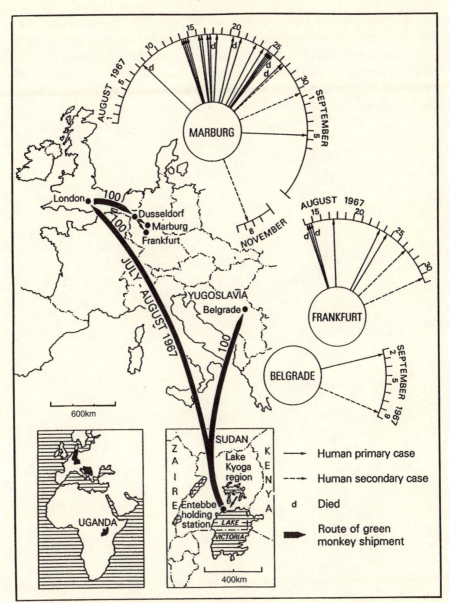

Fig. 3.12. History of the 1967 Marburg (West Germany) fever outbreak in Europe. The vectors show the routes of green monkey shipments from Uganda. The three city diagrams show the timing of primary and secondary cases with deaths (d) indicated.

Source: Smallman-Raynor *et al.* 1992, *op. cit.* [note 6], Fig. 3.4A, p. 133.

disease, after the city from which it was first reported, symbolizes the ways in which twentieth-century changes in transport have allowed the rapid spread of epidemics. The slow movements traced by Sauer in the prehistoric period have now been replaced by one in which which no part of the globe cannot be reached by vectors within a few hours. We turn to the control implications of this new situation in the final lecture.

4

Containing Epidemic Spread

If the first lecture was programmatic, the second regional, and the third speculative then this final lecture represents a still further shift of focus. For I now turn to an immensely practical matter, that of containing epidemics. Here, the essential question is pragmatic: How do we delay or stop epidemic waves spreading in geographical space?

First, I set out a fourfold schema for the different spatial and non-spatial control strategies. Then, examples of each of the four types is considered in turn, culminating in an examination of the prime achievement of international public health during the twentieth century—the global eradication of smallpox. We ask how far smallpox provides a model for other communicable diseases. In conclusion, we note briefly some of the problems that are likely to effect disease control in the next few decades.[1]

Spatial control strategies

It is helpful to consider the problem of control strategies for communicable diseases by setting it within a modelling framework. We look here at a very simple model of disease transmission touched on in the first lecture, and consider its implications for endemic reservoirs and how it may be translated into general control and forecasting frameworks.

The ways in which diseases are transmitted has attracted mathematical interest from Bernoulli onwards: the classic account is given by Bailey in his *Mathematical Theory of Infectious Diseases*.[2] To give a flavour of this approach, a very simplified diagram of the spread of measles infection through a human population is illustrated in Figure 4.1. This shows that we need to maintain an unbroken chain of infectives if an epidemic is to be generated and sustained.

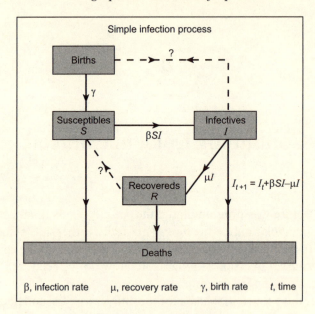

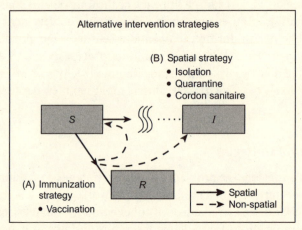

Fig. 4.1. Simplified model of control strategies for an infection process. (*Above*) Main components in the model based on the Hamer–Soper model of measles spread. (*Below*) Alternative intervention strategies: spatial intervention based on (B) blocking links between infectives and susceptibles and (A) opening of new pathways through immunization that outflanks the infectives (*I*) box.

Source: Cliff *et al.* 1993, *op. cit.* [note 1], Fig. 16.1, p. 414.

Protection against the spread of infection can be taken at two points in this flow diagram. First, the route from susceptibles to recovereds can be short-circuited by the establishment of immunity through some variant of immunization, as reviewed historically by Hinman and by Spink.[3] A second method is to prevent or interrupt the mixing of infectives and susceptibles by erecting protective spatial barriers. This may take the form of isolating an individual or a whole community, or restricting the geographical movements of infected individuals by quarantine requirements; another approach is by locating populations in supposedly safe areas. For animal populations, there exists a third possibility: the creation of a *cordon sanitaire* by the wholesale evacuation of areas or by the destruction of those infected.

If we look at the strategy by which smallpox was finally eradicated (see later in this lecture), we can see the control process as one of the progressive spatial reduction in the areas of the world in which the disease was endemic. Cliff and Haggett have presented a stage-by-stage schema of possible reduction strategies, stressing the ways in which geographical considerations impinge on control by vaccination.[4] These different spatial control strategies are illustrated in Figure 4.2; in each of the four diagrams, infected areas are stippled, while disease-free areas are left blank.

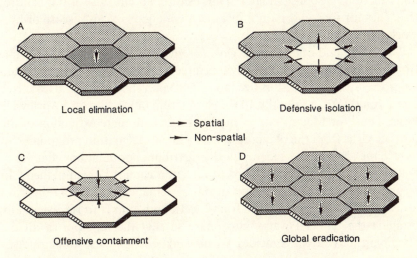

Fig. 4.2. Schematic diagram of four spatial and non-spatial control strategies to prevent epidemic spread. Infected areas are stippled; disease-free areas left white. Geographical areas shown arbitrarily as hexagons.
Source: Cliff and Haggett, 1989, *op. cit.* [note 1], Fig. 2, p. 318.

In the first stage, *local elimination*, the emphasis is on breaking, in some particular location, the disease chain by vaccination. The programmes aimed at eliminating indigenous measles in countries such as the United States, Czechoslovakia, and Australia, illustrate this phase. Such vaccination programmes may themselves be variable within a country and have a geographical component. That is, priority may be given to certain groups (e.g. the elderly) which have themselves a distinctive geographical distribution.

Once an area is cleared of an indigenous disease, then there is a need for a second stage, *defensive isolation*, which entails the building of a spatial barrier around a disease-free area. Attempts to erect such barriers (notably at sea ports) were made in the nineteenth century but may be impractical today. We have already noted in concluding the last lecture the difficulties that air travel poses to the use of quarantine to prevent infectious cases from gaining access to susceptible populations; the United States experience with its measles elimination programme illustrates the point.

A third stage, *offensive containment*, is a more appropriate approach in these circumstances. This is the reverse of the second case in that the occasional spread of a local outbreak within a larger and generally disease-free area is halted and progressively eliminated by a combination of vaccination and isolation. We consider later in this lecture, Tinline's use of ring-control strategies for foot-and-mouth disease, a contagious virus disease of livestock.[5]

The fourth and final stage of *global eradication* would arise in principle from the combination of the previous three methods: infected areas would be progressively reduced in size, and the coalescence of such disease-free areas would lead, eventually, to the elimination of measles on a worldwide basis. Thus, Sir Macfarlane Burnet's vision of ultimate extinction of the measles virus from the planet rests on a global vaccination programme to reduce the sizes of the geographically distributed populations that are at risk to levels at which the chains of infection cannot be maintained.[6] In terms of the Bartlett model, this means systematically reducing the wave order of different communities from I to II, and from II to III, eventually bringing the type III waves into phase so that the fade-out of all the remaining active areas coincides. We now take each of these four strategies in turn.

Local elimination and natural breaks in infection

To understand local elimination of an infectious disease it is important to recall that in many infectious diseases, temporal and spatial breaks in the disease chains occur naturally. We noted in the opening lecture that the survival of the causative agent of an infectious disease is, *inter alia*, a function of the size of the population in which it is present. For a disease to be endemic, enough individuals at risk (susceptibles) must be present at all times for the chain of transmission of the agent from individual to individual to remain unbroken.

For the infectious disease most commonly used in modelling studies (measles), the critical community size required to sustain endemicity has been studied in detail in two classic papers by Bartlett and Black.[7] Bartlett plotted the mean period (time interval) between epidemics (in weeks) for a sample of nineteen English towns (recall Figure 1.8). The time interval was found to be inversely related to the population size of the community. Given the reporting rates for measles at the time, this implied that a population of around 250,000 to 300,000 is required to ensure continuous transmission chains of infection.

Black extended Bartlett's work by examining the relationship between measles endemicity and population size in eighteen island communities (recall Figure 2.1). Of the islands studied by Black, only Hawaii with a total population (then) of 550,000 displayed clear endemicity. Other islands close to Bartlett's 250,000 value just failed to display endemicity. This may, of course, reflect the difference between the isolation of islands as opposed to the mainland location of Bartlett's cities.

The basic notion of a threshold population, below which an infectious disease becomes naturally self-extinguishing, is paramount in articulating control strategies.[8] It implies that vaccination may be employed to reduce the susceptible population below some critical mass so that biological processes may achieve the rest. As a result, attempts have been made to establish the endemicity thresholds for a variety of transmissible diseases. Once the population size of an area falls below the threshold then, when the disease concerned is eventually extinguished, it can only recur by reintroduction from other reservoir areas.

The impact of vaccination on epidemic cycles

The impact of partial vaccination policies upon the size and spacing of recurrent epidemics has attracted considerable attention.[9] Working with measles, Griffiths has used the Hamer–Soper and chain binomial models to examine the long-term effect on a community of a continuing partial vaccination programme. If x denotes the proportion of children not artificially immunized by vaccination, Griffiths found that the critical community size for endemicity is multipled by $1/x^2$. Thus, 50 per cent immunization increases the critical community size from 250,000 to 1 million, whereas 90 per cent immunization increases the threshold to 25 million. The impact of partial vaccination policies on the size and spacing of recurrent epidemics has been considered in a series of mathematical investigations.

Figure 4.3 shows the predicted effect of partial immunization, sustained over fifteen years, at 80 per cent of the 1- to 2-year-olds in a theoretical population. The slow damping of epidemic amplitude is evident as the cumulative impact of vaccination is felt and, eventually, the endemic cycle is broken and whole epidemics are missed. Thus, natural fade-out will become very widespread, enhancing the possibility of local elimination and eventually global eradication. Theoretical studies of such immunization strategies have been widely conducted.[10]

Figure 4.4 shows the use of an epidemiological control model by Anderson and Nokes to predict the numbers of cases of another major infectious disease, mumps.[11] Their results show mumps infection varying across age classes and through time, before and after the introduction of a programme of mass cohort immunization. (In the model the vaccination is assumed to reach 60 per cent of all 2-year-olds with an average age at infection before immunization of 6.7 years.)

In the pre-vaccine period, the epidemic peaks show the majority of mumps cases occurring in the youngest age classes. Following vaccination, there is an obvious and expected decline in infection incidence (particularly in the young). But the diagram also shows that the age at which the *remaining* cases of mumps occur now goes up. This is marked by the wave of infections migrating, in time, into the older age classes. The implications of such shifts in the age distribution of cases may have important clinical consequences in some diseases.

One of the best examples of a sustained programme aimed at local elimination within a country is provided by the United States measles camapign. In that country, in the early years of the twentieth century, thou-

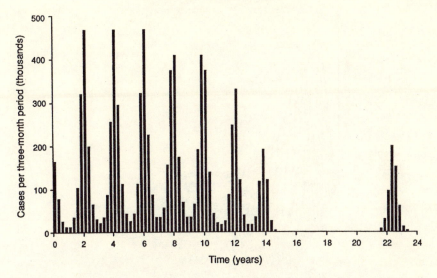

Fig. 4.3. Predicted effects of widespread immunization. Applications of the Anderson and May model with the level of immunization held constant for 15 years at 80 per cent of 1- to 2-year-olds.
Source: Cutts and Smith, 1994, *op. cit.* [note 29], Fig. 10, p. 23.

sands of deaths were caused by measles each year and, at mid-century, an annual average of more than half a million measles cases and nearly 500 deaths were reported in the decade from 1950 to 1959. It was against this background that the main federal health control body, the Centers for Disease Control, Atlanta, Georgia (CDC) evolved in the United States a programme for the elimination of indigenous measles once a safe and effective vaccine was licensed for use in 1963.[12] As we noted earlier in this lecture, it is estimated that a population of the order of 250,000–300,000 is required to maintain endemic measles. Work in Africa by McDonald in the early 1960s led him to suggest that one way of reducing the 'at-risk' population in large countries below this endemicity threshold was by mass vaccination, so breaking the chains of measles infection. In the countries studied by McDonald, he argued that an annual mass vaccination campaign reaching at least 90 per cent of the susceptible children would have the required effect.

In 1966, CDC announced that the epidemiological basis existed for the eradication of measles from the United States using a programme with four tactical elements: (1) routine immunization of infants at one year of age; (2)

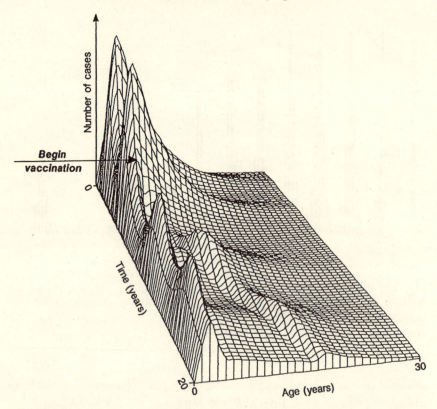

Fig. 4.4. Time–age predictions of the number of cases of a virus disease following vaccination. Numbers of mumps cases (vertical axis) are plotted against time (in years) and age (in years). Vaccination is by mass cohort immunization (60 per cent of 2-year-olds).
Source: Anderson and Nokes, 1991, *op. cit.* [note 11], Fig. 14.4c, p. 250.

immunization at school entry of children not previously immunized (so-called 'catch-up' immunization); (3) surveillance; and (4) epidemic control. The immunization target aimed for was 90–95 per cent of the childhood population.

Following the announcement of possible measles elimination, considerable effort was put into mass measles immunization programmes throughout the United States. Federal funds were appropriated and, over the next three years, an estimated 19.5 million doses of vaccine were administered. The discontinuity induced in the time series of reported cases is shown for the United States as a whole in Figure 4.5(A, B). In 1962, the year before the measles vaccine was introduced, there were 481,500 cases of measles reported in the United States. By 1966, this number had been reduced by more than 50 per cent to 204,000 and, by 1968, the reported incidence had plummeted to 22,000, less than 5 per cent of the 1962 level. But in 1969, a vaccine against rubella (German measles) was licensed and all federal funds were targeted against rubella; no federal funds were allocated to the measles immunization programme from 1969 to 1971. As a result, public sector vaccination declined. The susceptible population rose and, as Figure 4.5 bears witness, the number of reported measles cases rose sharply, reaching 75,000 cases in 1971.

By the mid 1970s, it was evident that the campaign against measles was running out of steam and that steady increases in incidence were occurring. To remedy this situation, a nationwide childhood immunization initiative was launched in April 1977, followed by the announcement in October 1978 of a programme to eliminate indigenous measles from the United States by October 1982. The immunization goal aimed for was again McDonald's 90 per cent of the childhood population.

The geographical impact of this second push against the disease is seen in the two maps in Figure 4.5(C, D). The maps show the distribution of counties in the United States reporting measles cases at the start of the campaign (1978) and five years later in 1983. The contraction of infected counties from most of the settled parts of the United States in 1978 to restricted areas of the Pacific Northwest, California, Florida, the north-eastern seaboard, and parts of the Midwest is pronounced. The persistence of indigenous measles in many of these regions may be explained by the importation of cases from Mexico (often through the illegal immigration of agricultural workers) and, to a much lesser extent, from Canada. In 1983, twelve states and the District of Columbia reported no measles cases, and twenty-six states and the district of Columbia reported no indigenous cases. Four states (Indiana, 406; Illinois, 216; California, 181; Florida, 159) accounted for 64 per cent of the 1,497 cases. Of the 3,139 counties only 168 (5 per cent) reported any measles cases. In contrast, measles was reported from 195 counties in 1982 and from 988 in 1978 when the Measles Elimination Programme began.

Unfortunately, total elimination in the United States still has not been achieved. Vaccination levels have fallen back and the continued importation of measles cases from overseas has resulted in a resurgence of cases to 6,200 in 1986. But note the way in which a neurological complication attributed to measles, subacute sclerosing panencephalitis (SSPE), has paralleled the decline in measles occurrence.

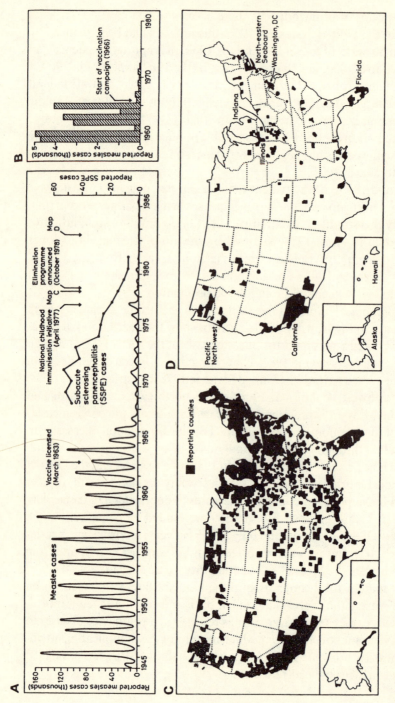

Fig. 4.5. Measles reduction in the United States. (A, B) Monthly measles cases 1945–86 with annual figures for SSPE (subacute sclerosing panencephalitis) cases for 1968–81 superimposed. (C, D) US counties reporting measles during one or more weeks in 1978 (C) and 1983 (D).

Source: Cliff and Haggett, 1988, *op. cit.* [note 1], Fig. 4.9, pp. 164–5.

As the general tide of measles cases has been rolled back in the United States, so small geographical pockets of susceptibles have been revealed. Figure 4.6 illustrates one such example of an outbreak amongst a religious group, the Amish, who have low immune status because they object to vaccination on theological grounds. The Amish trace their religious heritage back to sixteenth-century Switzerland and first moved to the United States in 1709 to escape persecution. Today, about a quarter live in rural

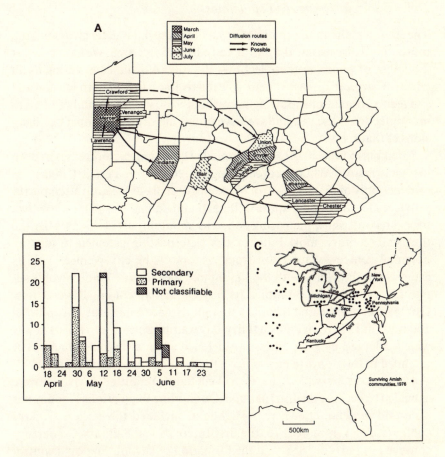

Fig. 4.6. Pennsylvania measles outbreak, 1987–8: spread in the Amish community. (*Above*) Dates of rash onset of the first Amish case in each county; vectors identify diffusion routes. Inset graph shows, by exposure status, date of rash onset among Amish patients in Lebanon County. (*Below*) Spread to Amish communities in other states.

Source: data from Sutter *et al.*, 1991 *op. cit.* [note 13], Fig. 1, p. 13.

Pennsylvania and the maps show how a major measles outbreak in 1987–8 was channelled through Amish communities both within and outside that state. Isolated groups with low vaccination rates behave like Bartlett C communities and are vulnerable to severe outbreaks often separated by long intervals of time.[13]

Defensive isolation against epidemics

The second plank in the spatial schema in Figure 4.2 is defensive isolation, preventing a disease invading an already disease-free area. We have noted at a number of points in these lectures, the role of distance in acting as an effective barrer to disease spread. The relative isolation of both Iceland and Fiji led to populations growing up without exposure to a number of common infectious diseases and we have noted the dramatic impact of invasions of that space.

Attempts to prevent the spatial spread of communicable diseases lie deep in human history. While the archaeological record from Peru to China gives evidence only of disease control by extensive civil engineering attempts (to provide safe water and dispose of human and animal wastes) the later historical records show increasing concerns about imported diseases. By the thirteenth century, most Italian cities were posting gatemen to identify potential sources of infection from visitors to the city. Venice, with its extensive trading links with the Levant and the Oriental lands beyond, pioneered the idea of quarantine. Its tiny Dalmatian colony, Ragusa (now Dubrovnik) saw the first recorded attempt to place a moratorium on travel and trade in 1377. Originally a thirty-day waiting period (a *trentino*), it was widely adapted by port cities as a defence against the plague and later extended to forty days (a *quarantino*), the familiar quarantine period.[14]

The process by which the early Venetian quarantine measures were extended to become the International Quarantine Regulations of today has been told in detail elsewhere.[15] Highlights included the earliest sanitary conference at Venice in 1576, the first International Sanitary Conference convened in Paris in 1851, and the formation of key international control bodies: the International Office of Public Hygiene in 1907, the Pan American Sanitary Bureau in 1910, the Health Section of the League of Nations in 1922, and the World Health Organization in 1946. For the medieval world the critical threat came from imported plague but by the nineteenth century it was concerns about cholera and yellow fever that drove the need for regulations. Smallpox, louse-borne typhus, and relapsing fever were not added to the international list of regulated diseases until 1922.

Of all the island communities that have adopted strict quarantine measures, Australia has proved of special interest. The very low population density of the indigenous Aboriginal peoples, the long sea journey to Britain after European settlement began in 1788, and the separation of settlements into 'limpet' colonies around the coast of the continent were all factors in providing a natural barrier of isolation. When infectious diseases that were common in Europe eventually landed in Australia they often formed very sharply peaked and discrete waves as the Kendall model (see Figure 1.10) predicted. The fact that rubella was identified in New South Wales as a sharp peak of cases rather than a background low incidence as in Europe was a critical factor in allowing Gregg and Lancaster to identify the dangerous effect of the virus on mothers in the first few months of pregnancy. The special role of natural isolation in acting as an epidemiological barrier has been studied by Oliver Lancaster in a series of pioneering studies.[16] (See Figure 4.7.)

Offensive containment

The third spatial case in Figure 4.2 is the inverse of the second. In this category, we assume a country or other geographical area to be normally free of a disease but occasionally outbreaks occur either at its borders or deeper within the territory. The problem is how to contain the invasion.

If an invasion occurs, it is essential to know (or try to know) how it is likely to spread. Spatial forecasting models using mathematical formulations to estimate the likely spread of an infectious disease are increasingly being used in epidemiological studies. Predictive models based on studies of epidemic waves tend to be of two kinds: predictions based on apparent regularities in the spatial and temporal records of a disease (*space–time models*); and predictions based on models of the epidemiological processes involved (*process models*). Regional examples were given in the second lecture on work by Cliff and colleagues on predicting measles waves in Iceland. Another good example is the work by Murray's team on predicting rabies waves in Britain.

Rabies, an acute virus disease that attacks the central nervous system, has attracted special attention from spatial modellers.[17] One major rabies

A   NEW SOUTH WALES CENSUS 1911, 1921, 1933

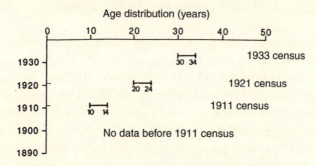

B   INSTITUTION FOR DEAF, DUMB, BLIND, DARLINGTON (NSW)

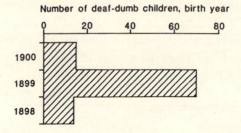

C   INSTITUTION FOR DEAF, DUMB, BLIND, DARLINGTON (NSW)

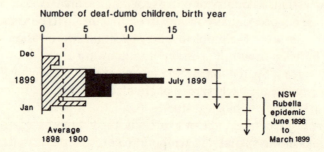

Fig. 4.7. Rubella (German measles) in New South Wales, Australia. (A) Position of cohorts in the New South Wales censuses of 1911, 1921, and 1933. (B) Number of deaf, dumb, and blind children in the Darlington institution, New South Wales, by birth month, 1898–1900, and (C) for 1899 in relation to rubella epidemic. *Source*: Cliff and Haggett, 1989, *op. cit.* [note 1], Fig. 8, p. 331.

epizootic has persisted and progressed steadily across Europe in a wave-like form for some forty years.[18] The epizootic started on the German/Polish border and the wave front has moved westwards at an average speed of 30–60 km per year, so that today it has nearly reached the English Channel coast. The main carrier is the fox and, because of their central role in prop-agating the epizootic, fox ecology has been intensively studied. Detailed work by MacDonald on the foraging range of vixens about the earths shows that the frequency distribution of distances travelled (the contact distribution) is accurately described by a negative exponential curve; dog foxes are more nomadic.[19] The shape implies that most journeys are within a short range of the earth and only a few are in excess of 5 kilometres. The theoretical properties of spread processes which have negative exponential contact distribution have been investigated by Mollison's group.[20] They have shown that, when the negative exponential rule is obeyed, wave prop-agation across geographical space will be of relatively uniform velocity and the wavefront will remain intact.

Murray's Oxford group has modelled the geographical variation in rabid and non-rabid fox density as an epizootic passes and give estimates of the speed of travel of the wave.[21] They suggest that a rabies wave will take between six and twelve years to travel 200–600 km so that, at a fixed loca-tion in the enzootic zone, waves of rabies cases will recur in the fox popula-tion every six to twelve years. Figure 4.8 shows model estimates of the time taken for a rabies epizootic introduced into the United Kingdom in the vicinity of Southampton to reach different parts of the country.

They assume that the geographical distribution of foxes is that given in MacDonald.[22] The map illustrates how wave speed depends upon the sus-ceptible fox density. Native fox density is lower in Wales, East Anglia, and Scotland than elsewhere in the United Kingdom and it is evident that the first two stand out as relatively 'late' islands bypassed by the main wave-front.

Another case study which illustrates the potential for modelling global transmission is provided by the work of Gould on the geographical spread of AIDS.[23] He took the largest 102 urban centres in the coterminous United States and used air passenger origin–destination data to compute a weighted 102×102 transition probability matrix. Probabilities are partic-ularly high amongst the five largest cities which in 1992 'exchanged' by air travel some 13 million people every year; likewise, probabilities are low amongst smaller and distant centres with small volumes of population exchange. In mathematical terms, the matrix forms an operator capable of

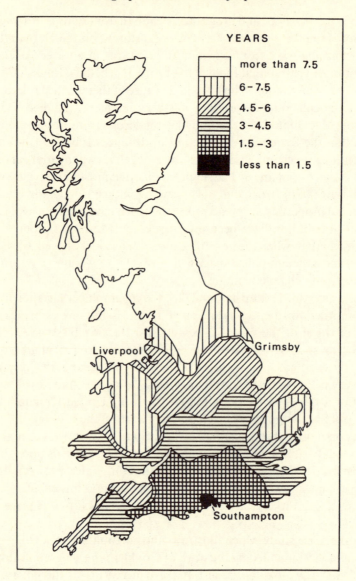

Fig. 4.8. Predicted map of rabies spread. Murray's estimate of the probable position of rabies front within Britain from a hypothetical introduction of the disease at the port of Southampton.

Source: Kallen *et al.*, 1985 *op. cit.* [note 21], Fig. 1, p. 378.

multiplying a state vector. The state vector in Gould's case was the distribution of AIDS cases on a city-by-city basis at a particular year (1986). After a series of probabilistic multiplications the 'projected' AIDS distribution for 1990 was calculated.

The results of the projections are shown in Figure 4.9. In the upper part of the figure (A), the projected values from the Gould model (horizontal axis) show a close approximation to the observed AIDS rates (vertical axis) with a correlation of about 80 per cent. The lower part (B) shows the 'residuals' from the projection (i.e. those cities under- and overpredicted by the model). Many of the negative residuals (towns with AIDS rates overpredicted) lie in the older 'rustbelt' towns of the north-east United States with many blue-collar workers of Catholic and recent immigrant backgrounds. Many southern cities that are tourist destinations come out as positive residuals: they have AIDS rates higher than the model would predict.

Models of this kind, where they can be calibrated on 'old' data from recent epidemic events, can be used to predict the spread of communicable diseases in the future. Russian workers have used such models on a world network of cities to project the spread of a new strain of influenza.[24] Cliff and Haggett have used similar models to predict the likelihood of measles imports into the several regions of the United States.[25] The construction of global early warning systems for the transmission of communicable diseases are in prospect.

Ring control strategies

A third stage in Hägerstrand's overall model (see Figure 1.12) relates to control policy. The central question is how the spread of an epidemic wave can be controlled through countermeasures, such as quarantine, isolation (e.g. school closure), or vaccination. Examples of non-spatial and spatial control policies are given (see Figure 4.2C).

Containing and eliminating an outbreak requires that some estimate be made of the rate of the geographical spread of the disease. If the estimate is too low and protective measures (e.g. vaccination) are concentrated in too narrow a defensive zone, then the outbreak will jump over the barrier. If the estimate is too high, the available resources for containment will be spread thinly over too large an area and the chances of success reduced. Clearly, there are some analogies with fire-fighting strategies for containing bush fires.

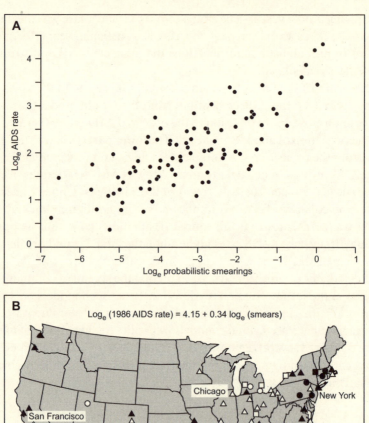

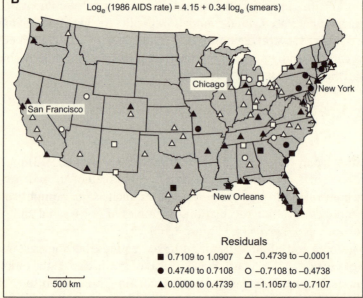

Fig. 4.9. Projective modelling of the AIDS epidemic in the United States. (A) Actual AIDS rates for 102 US cities on vertical axis plotted against the probabilistic estimates on the horizontal axis. (B) Maps of residuals to show which cities were 'overpredicted' and which 'underpredicted' by the model.

Source: Gould, 1995, *op. cit.* [note 23], p. 27.

Foot-and-mouth disease (FMD) is a virus disease of cloven-footed animals (cattle, sheep, pigs, and goats). Since 1892, control in the United Kingdom has been exercised through total slaughter of all exposed herds, because the virus is so readily transmitted among cattle. Even if an animal survives, the economic value of an infected animal is much reduced. The aim of this policy is to destroy all possible reservoirs of infection and thereby eliminate the disease.

In the late 1960s I was fortunate to have a Canadian doctoral student at Bristol, who was interested in extending the Hägerstrand model to epidemic control situations. An opportunity arose in the aftermath of the great 1967–8 FMD epizootic in Britain. Tinline explored the use of geographically targeted vaccination strategies, in conjunction with slaughter, as a more cost-effective way of containing FMD.[26] He was able to demonstrate, using data from the 1967–8 outbreak, that the airborne spread of FMD virus (which is one of the smallest viruses known) downwind from sources of infection was a major cause of additional outbreaks. By the time the disease was positively identified in a core area, and slaughter carried out, virus particles had often been carried long distances by the wind, subsequently to be redeposited, after rain, beyond the FMD slaughter area.

Tinline therefore suggested a scheme of *ring vaccination* in areas downwind of initial outbreaks to contain subsequent spread. The principles involved are illustrated in Figure 4.10. In the first map of this figure (4.10A), blanket vaccination of all herds downwind of the outbreak is called for on day 1 after confirmation (all cells coded 1). However, this is generally impossible in any major outbreak because of shortage of vaccination teams. Vaccination out to a distance of 20 kilometres from an outbreak was found to be necessary to construct a *cordon sanitaire* to contain the disease.

To recognize veterinary manpower limitations, Tinline evaluated three ring vaccination schemes. These are shown in the remainder of Figure 4.10(A). Cells are coded 1 and 2 according to the priority for vaccination after confirmation of FMD (0 = no vaccination). Figure 4.10(B) shows the observed number and percentage of FMD deaths with no vaccination, blanket vaccination in all areas, and for each of the ring vaccination schemes. Apart from blanket vaccination, the most successful is scheme III. The critical feature is vaccination from the outside–in towards the centre of the outbreak. Tinline found that blanket vaccination was called for in a zone 13–20 km from the outbreak within two days of confirmation of the disease; the ring 7–13 km from the outbreak needed to be treated in days 3 and 4, and the core zone on days 5 and 6. It was also found that the lower

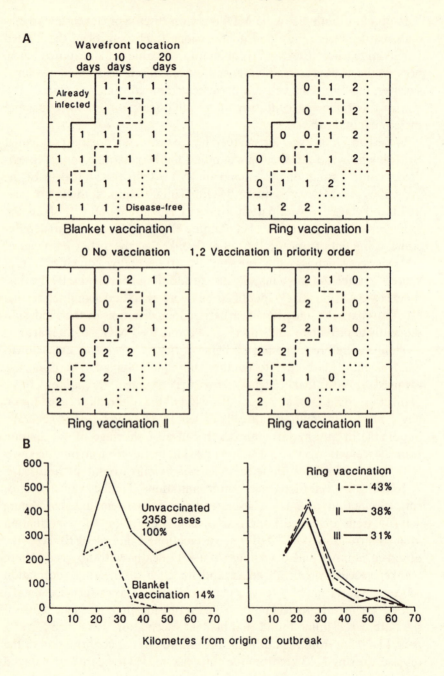

A

Wavefront location
0   10   20
days days days

Blanket vaccination

Ring vaccination I

0 No vaccination   1,2 Vaccination in priority order

Ring vaccination II

Ring vaccination III

B

Unvaccinated
2,358 cases
100%

Blanket
vaccination 14%

Ring vaccination
I ---- 43%
II —— 38%
III —— 31%

Kilometres from origin of outbreak

the vaccination cover and the slower the administration of the vaccine, the greater was the benefit of the outside–in approach.

The practical difficulty in implementing ring vaccination policy is the need to forecast accurately wind conditions in the vicinity of an outbreak, since these will dictate both the location and geographical extent of the ring. The presence of lee waves will further complicate affairs. Given the need to mobilize vaccination teams and the length of time required for vac- cination to induce immunity in animals, wind forecasts for the next twenty days after an outbreak were required for optimal results. In practice, com- pulsory ring housing of animals would have also dramatically reduced exposure to the airborne virus.

Global eradication: the smallpox campaigns

Finally, we turn to the fourth case of epidemic containment in Figure 4.2, the global eradication of a disease. When the epidemiological history of the twentieth century comes to be written, the outstanding success that histori- ans will be able to record is the global eradication of smallpox. The complex story which culminated in the last recorded natural case in October 1977 (there were to be two subsequent laboratory deaths) has been superbly told and in massive detail by Fenner and can only be summarized here.[27] That success has inevitably raised questions as to whether other infectious dis- eases, measles among them, can also be eradicated. In this section, we look briefly at the smallpox eradication programme and compare its success with the prospects for two other diseases—measles and poliomyelitis. We follow Fenner in confining the term *eradication* to the total elimination of the infectious agent (except, as with smallpox, for preserved laboratory examples); elimination refers to stamping out the disease in a particular country or region but leaves open the possibility of reinfection from another part of the world.

Fig. 4.10. Spatial impact of disease control strategies. Tinline's simulations of dif- ferent vaccination policies against the spread of the 1967–8 foot-and-mouth (FMD) epizootic in central England. (A) Blanket vaccination and three different ring-control vaccination strategies. Each cell on the map covers a 10 km × 10 km area. (B) Spatial cross-section of FMD cases plotted against distance for the con- trol strategies shown in (A).
Source: Tinline, 1972, *op. cit.* [note 5], Fig. 6.17, Table 6.12.

Although the World Health Organization (WHO) has, from time to time, conducted major campaigns against infectious disease (notably malaria and yaws) only one disease—smallpox—has so far been globally eradicated. The practical reality of devising, coordinating, and financing a field programme involving more than thirty national governments and some of the world's most complex cultures and demanding environments proved to be of heroic proportions.

Until the mid 1960s, control of smallpox was based primarily on mass vaccination to break the chain of transmission between infected and susceptible individuals by eliminating susceptible hosts. Although this approach had driven the disease from the developed world, the lesser developed world remained a reservoir area. Thus, between 1962 and 1966, some 500 million people in India were vaccinated, but the disease continued to spread. Between 5 and 10 per cent of the population always escaped the vaccination drives, concentrated especially in the vulnerable under-15 age group. Nevertheless, the susceptibility of the virus to concerted action had been demonstrated and led to critical decisions at the Nineteenth World Health Assembly in 1966.

This assembly embarked upon a ten-year global smallpox eradication programme which was launched in 1967. It started with mass vaccination, but rapidly recognized the importance of selective control. Contacts of smallpox cases were traced and vaccinated, as well as the other individuals in those locations where the cases occurred.

The programme had four main phases in each area targeted. In the *preparatory phase*, before active eradication was started, time was allowed for the epidemiological assessment of the distribution of smallpox and immunity in the local population. For example, epidemiological assessment was organized in India by geographical areas. In each area, the task was to record the location of each village, and whether or not evidence of smallpox was to be found in the village population. During the preparatory phase, health care personnel were also recruited and trained. Education programmes were established to ensure acceptance of vaccination.

The *attack phase* (while the incidence of smallpox in the targeted community was five or more cases per 100,000 population per annum, and less than 80 per cent of the population was vaccinated) consisted of systematic mass vaccination and the establishment of a surveillance programme with follow-up vaccination of contacts and individuals in local areas where cases occurred.

The *consolidation phase* was reached when smallpox incidence fell below five cases per 100,000 population per annum and primary vaccination had extended to over 80 per cent of the population. This phase consisted of a *maintenance vaccination programme* for newborns and those missed, such as immigrants, in the attack phase. The surveillance network now became critical, with every suspected case followed up by field investigation and action where necessary.

The *maintenance phase* was reached when there was no endemic small-pox in the targeted area for more than two years, while the disease still persisted on the continent concerned. Maintenance vaccination was continued and intense surveillance maintained. Each report of a suspected case was treated as an emergency until the final elimination occurred.

The success of this four-phase programme after 1966, following on the efforts of individual nations in the post-war period, may be judged from the maps and graphs in Figure 4.11.[28] By 1970, retreat was in progress in Africa. By 1973, the disease had been eliminated in Latin America and the Philippines; a few strongholds remained in Africa, but most of the Indian subcontinent remained infected. Despite a major flare-up of the disease in 1973 and 1974, the hunt by WHO for cases and case contacts continued. By 1976, the disease had been eradicated in South-east Asia and only a part of East Africa remained to be cleared.

The world's last recorded smallpox case was a 23-year-old man of Merka town, Somalia, on 26 October 1977. After a two-year period during which no other cases (other than the laboratory accident) were recorded, WHO formally announced at the end of 1979 that the global eradication of small-pox was complete.

Global campaigns for other epidemic diseases

The dramatic success of the WHO smallpox programme has inevitably raised the prospect and hope that other virus-borne diseases can be eradicated. In 1974, WHO established its Expanded Programme on Immunization (EPI) with the objective of greatly reducing the incidence of six other crippling diseases: diphtheria, measles, neonatal tetanus, pertussis, poliomyelitis, and tuberculosis. Two other diseases were later added. Table 4.1 summarizes some of the characteristics of the disease and indicates in the final two columns the continental variation in the vaccination levels achieved to date. Comprehensive reviews of the role of such vaccination programmes on world health are given in Cutts and Smith's *Vaccination and World Health* and in the WHO *Immunization Policy*.[29]

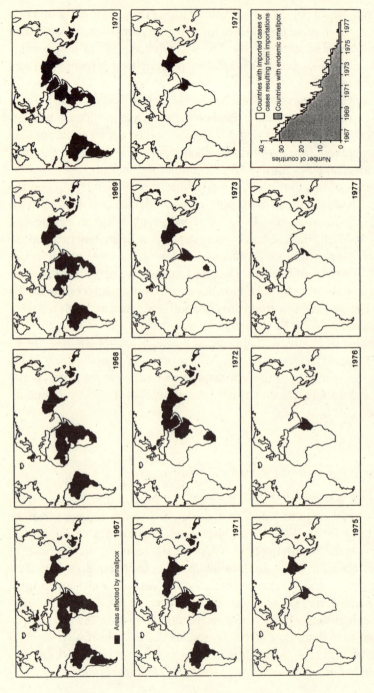

Fig. 4.11. Global eradication of smallpox under the WHO Intensified Programme, 1967–77. Countries with smallpox cases in the year in question shown in black.

Source: Redrawn from maps and graphs in Fenner *et al.* 1988, *op. cit.* [note 27], Fig. 10.4, Plates 10.42–10.51, pp. 516–37, *passim.*

Table 4.1. Target diseases in the WHO expanded programme on immunization

Disease	Infectious agent	Reservoir	Spread	Nature of vaccine	Form of vaccine (No. doses)	Immunization coverage (%) Africa	Immunization coverage (%) Europe
Diphtheria	Toxin-producing bacterium (*Corynebacterium diptheriae*)	Humans	Close contact respiratory or cutaneous	Toxoid	Fluid (1)	50	86
Hepatitis B	Virus	Humans	Perinatal Child–child Blood Sexual	HB vaccine	Fluid (3)	0.15	12
Measles	Virus	Humans	Close respiratory contact and aerosolized droplets	Attenuated live virus	Freeze-dried (1)	49	78
Pertussis	Bacterium (*Bordetella pertussis*)	Humans	Close respiratory contact	Killed whole-cell pertussis bacterium	Fluid (3)	50	86
Poliomyelitis	Virus (serotypes 1, 2, and 3)	Humans	Faecal–oral Close respiratory contact	Attenuated live viruses of 3 types	Fluid (4)	50	92
Tetanus	Toxin-producing bacetrium (*Clostridium tetani*)	Animal intestines Soil	Spores enter body through wounds, umbilical cord	Toxoid	Fluid (3)	35	na
Tuberculosis	*Mycobacterium tuberculosis*	Humans	Airborne droplet nuclei from sputum-positive person	Attenuated *M. Bovis*	Freeze-dried (1)	68	81
Yellow fever	Virus	Humans Monkeys	Mosquito-borne	Attenuated live virus	Freeze-dried (1)	6	na

Note. Immunization coverage at March 1994. Africa excludes South Africa. HB vaccine, hepatitis B vaccine; na, not applicable.
Source: Based on data provided by the Global Programme for Vaccines and Immunization. World Health Organization, 1995, *op. cit.* [note 29], Tables 1–3, pp. 2–5.

It is important to consider how far smallpox was a useful control model for other communicable diseases. For whatever the huge difficulties in practice, in principle, smallpox was well suited (perhaps uniquely well suited) to global eradication: Fenner has summarized the special characteristics of smallpox which allowed global eradication (Table 4.2), and we follow his account.[30]

Table 4.2. Biological and sociopolitical features that favoured the global eradication of smallpox

---

Biological features

1. A severe disease, with high mortality and serious after-effects.
2. No animal reservoir of variola virus.
3. Very few subclinical cases.
4. Cases became infectious at the time of onset of rash.
5. Recurrence of infectivity never occurred.
6. Only one serotype existed.
7. An effective, stable vaccine was available.

Sociopolitical features

8. Earlier country-wide elimination showed that global eradication was an attainable objective.
9. There were few social or religious barriers to the recognition of cases.
10. The costs of quarantine and vaccination for travellers provided a strong financial incentive for wealthy countries to contribute.
11. The Intensified Smallpox Eradication Unit of the WHO had inspirational leaders and enlisted committed heath workers.

Source: Fenner, 1986, *op. cit.* [note 30], Table 1, p. 37.

First, smallpox was such a severe disease that it was clearly worth the effort required for eradication. Second, whatever its origins may have been, variola virus was a specific human virus; there was no animal reservoir. Third, subclinical infections were virtually unknown, and those that did occur excreted very little virus and were of no epidemiological importance. Fourth, spread usually resulted from direct face-to-face contact with patients with a rash; patients were not infectious during the incubation period or the pre-eruptive phase. If cases were isolated as soon as the rash was apparent, in a setting in which they had contact only with vaccinated or immune persons, the chain of transmission could be broken. Fifth, neither a prolonged carrier state nor recurrence of clinical illness with associated infectivity ever occurred in smallpox; hence the disappearance of acute infections meant that the chance of transmission had been eliminated. Sixth, there was only one serotype of variola virus—over centuries of time and all over the world—and, seventh, since the time of Jenner, we have had an effective live-virus vaccine against smallpox. In the 1950s, a freeze-dried vaccine was developed that was stable even under the most adverse conditions.

Fenner also recognizes that the biological features of smallpox, while a necessary precondition for global eradication, were not in themselves sufficient to ensure success. Several sociopolitical factors were also crucial (Table 4.2). Point 10 in the table argues that smallpox imposed a heavy financial burden on the industrialized countries, as well as on those where smallpox was endemic; the disease was economically significant in the West. Quite apart from the disease and death from smallpox itself, the cost of vaccination, plus that of maintaining quarantine barriers, is calculated to have been about $1,000 million per annum in the last years of the virus's existence in the wild. Fenner stresses that these costs disappear completely if, and only if, *global* eradication is achieved. Points 9 and 11 in the table are self-explanatory.

A detailed comparison of both the biological and socioeconomic characteristics of smallpox with those for measles and poliomyelitis shows only a partial overlap (Table 4.3): measles shares ten of the fourteen smallpox characteristics but poliomyelitis only five out of fourteen.

Table 4.3. Smallpox, measles, poliomyelitis: comparison of biological and sociopolitical features favouring global eradication.

Features	Poliomyelitis	Smallpox	Measles
Biological			
1. Reservoir host in wildlife	No	No	No
2. Persistent infection occurs	No	No	No
3. Number of serotypes	1	1	3
4. Antigenetic stability	Yes	Yes	Yes
5. Vaccine effective	Yes	Yes	Yes
Cold chain necessary	No	No	Yes
Number of doses	1	1	4
6. Infectivity during prodromal	No	Yes	Yes
7. Subclinical cases occur	No	No	Yes
8. Early containment of outbreak possible	Yes	No	No
Sociopolitical			
9. Country-wide elimination achieved	Yes; many countries	Yes; few countries	Yes: few countries
10. Incentive for industrialized countries to assist	Strong	Weak	Weak
11. Records of vaccination required	No	Yes	Yes
12. Improved sanitation required	No	No	Yes

Source: Fenner, 1986, *op. cit.* [note 30], Table 2, p. 39.

We noted earlier in the lecture that the United States' lead in measles elimination was followed up with varying degrees of vigour in other Western countries and pressed in developing countries through the EPI. But by 1995, the position achieved was spatially very patchy. A few countries (e.g. Hungary, Israel, and Sweden) claimed vaccination rates of over 95 per cent. In contrast, the average coverage for Africa was 51 per cent, Latin America 81 per cent, South East Asia 84 per cent. Worldwide, currently 45 million cases of measles occur each year; 1 to 2 million of them die (mostly young children) in developing countries. By 1993, the global coverage of measles vaccination was probably around 78 per cent, still well below the 90 per cent target originally set by WHO for 1995. Recent scares about MMR (combined measles, mumps, and rubella) vaccination and its hypothesized link to Crohn's disease and to autism in the late 1990s further reduces the likelihood of measles eradication being achieved in the near future.

It has become progressively clear that no matter how high coverage is, the infectivity of measles is likely to make global eradication a very long-term prospect. Even a vaccine coverage of 100 per cent will leave vaccine recipients susceptible unless the vaccine was more effective than the present 80-plus percentage level. The United States experience shows that vigilant efforts to maintain high vaccination levels, strong surveillance, and an aggressive response against imported cases in measles-free zones are required to hold the ground already gained. Any immediate hopes of the global eradication of measles seems remote but, for measles at least, a feasible scenario is that more developed countries will follow the lead of the United States and try to eliminate measles nationally as an endemic disease. For this to be achieved, however, the great divergence of attitudes to, and of programmes against, measles in the developed world will need to be unified. Whether the coalescence of disease-free zones in developed countries would ever allow a sustained attack on measles reservoirs in developing countries will depend as much on politics and economics as on epidemiology.

Poliomyelitis elimination campaigns

Eleven years after the close of its successful smallpox campaign, the forty-first World Health Assembly, meeting in Geneva in 1988, committed WHO to the global eradication of a second disease, poliomyelitis. Like smallpox, this target involves not only eliminating the disease, but totally eradicating the causative virus. The goal was made possible by forty years of research and vaccine development since Enders, Weller, and Robbins succeeded in growing poliovirus in cell culture. The licensing of the Salk inactivated (1955) and Sabin attenuated live vaccine (1961) was reinforced by the early success of the countries of the Pan American Health Organization which had agreed in 1985 to eradicate the wild poliovirus from the Americas.

The level of global success achieved by 1993 is shown in Figure 4.12.[31] The map shows that no countries in the Americas reported cases and that Europe, Japan, Australia, and New Zealand were free of cases. Tropical Africa and South and East Asia remained major zones where disease incidence remained high. Overall, the level of vaccination worldwide has risen from less than 5 per cent of children in 1974 to over 80 per cent in 1994. Over the same period the number of reported cases worldwide had fallen from a peak of over 70 million to less than 7 million. Since 1993, further contraction within Africa and Asia has been reported.

The World Health Organization has warned against complacency. Declining polio incidence mostly reflects individual protection from immunization, and not 'wild' virus eradication. For although surveillance is improving, less than 15 per cent of cases are being officially reported. The global strategy has five components: (1) high immunization coverage with oral polio vaccine, (2) sensitive disease surveillance detecting all suspected cases of poliomyelitis, (3) national or subnational immunization days, (4) rapid, expertly managed outbreak response when suspected cases are detected, and (5) 'mopping-up' immunization in selected high risk areas where 'wild' virus transmission may persist.

The major cost of eradicating poliomyelitis will be borne by the endemic countries themselves but donor country support will be required for vaccine, laboratories, personnel, and research. Of these, the most urgent need is for vaccine: although each dose of oral vaccine currently costs only 7 US cents, over 2 billlion doses will be required per year for routine and mass immunization. In the longer run, the economic benefits of disease eradication far exceed the cost. Since the last case in 1977, the United States has saved its total contribution once every 26 days. If present progress is maintained, the global initiative will start to pay for itself by the year 1998, producing savings of half a billion dollars by the beginning of this century increasing to $3 billion annually by the year 2015.

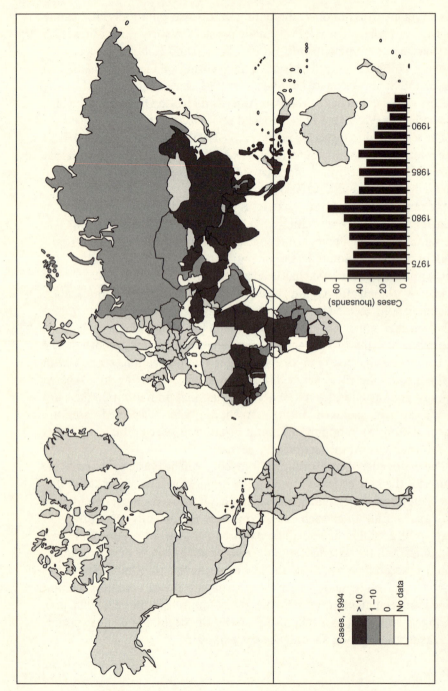

Fig. 4.12. Progress on the eradication of poliomyelitis. World incidence of indigenous poliomyelitis in 1994. (*Inset*) Annual number of cases of poliomyelitis notified in the world, 1974–93. *Source:* World Health Organization, 1995, *op. cit.* [note 31], p. 1.

Conclusions

This last lecture has only touched on the wide range of issues which surround the global control of communicable diseases. Looking forward into the future we see a series of trends which will influence control measures in the coming decades. While these will contain positive improvements in vaccine power and efficiency, these will be balanced by microbiological resistance. We see five contextual trends that will affect the ability to exert global control.

1. *Disease control is likely to rely less and less on conventional spatial barriers.* The speed of modern air transport (most of the world's cities are now within thirty-six hours of each other) and the complexity of air connections (there are now over 4,000 airports in the world with regular scheduled services and at least twice that number again for unscheduled services) make the traditional 'drawbridge' strategy increasingly irrelevant. The quarantine barriers first set up by the earliest International Sanitary Conferences were modelled to fit a slower mode of travel, notably ships, and fewer connection points.

2. *Rapid reporting and surveillance are likely to be increasingly critical in spatial control.* The Internet revolution is having an increasing impact on epidemic control. The use of electronic reporting (such as the United States National Electronic Telecommunications Surveillance System; NETTS) is likely to be extended worldwide through the Internet. Weekly bulletins such as the WHO *Weekly Epidemiological Record* or the CDC *Mortality and Morbidity Weekly Record* are now being regularly updated on an on-line basis.

3. *Ever-widening lists of communicable diseases and the high cost of surveillance will make sampling essential.* The legal requirements to notify critical infectious diseases is tending to be replaced by sampling systems in which sentinel practices are used to pick up trends in disease prevalence. This will intensify the legal problems associated with vaccination, identification, constraints on freedom of movement, and the increasing constraints which they pose.[32]

4. *Mathematical methods will increasingly supplement other epidemiological tools in global control.* In addition to the spatial models discussed earlier in this lecture, there will be the need to scan regularly the torrent of international and local data for 'aberrant' behaviour. Traditional CPE (current/previous experience) graphs will be

replaced by automatic monitoring in which anomalous events will be highlighted for the epidemiologist to consider.[33] Such anomalies which range from clinical reports of resistant malaria strains, to unusual clusters of meningitis cases, to higher than average influenza reports.

5. *Disease control and socioeconomic development are likely to be ever more closely tied together*. In his *World Health Report* for 1995, the Director-General of the WHO stated that: 'The world's most ruthless killer and the greatest cause of suffering on earth is listed in the latest edition of WHO's International Classification of Diseases under the code Z59.5. It stands for extreme poverty.'[34] The numbers of extreme poor have been rising over the last decade at a rate above gross population growth and in 1990 were estimated at over 1.1 billion, over one-fifth of humanity. For communicable diseases the links between poverty and disease come through many channels: absence of knowledge of protective measures, poor diet, lack of vaccination, clean water, and sanitation. The correlation runs the gamut of geographical scales from the global North–South contrast between the developed and developing worlds to the local contrast between the affluent suburbs and deprived inner-city ghettos of a Western city.

The above five are some of the more important contexts of change against which control is likely to be set. Each century, public health has had to fight disease with the tools available and the constraints imposed at the time. This century will be no exception as it prepares to fight both old diseases causing old problems, old diseases causing new problems (e.g. drug resistance), as well as wholly new diseases. There is growing evidence that a sensitivity to geographical inputs, whether in the form of new ways of mapping disease space or in recognizing significant regional variability in epidemic behaviour, will be an increasingly vital input into the complex multidisciplinary programmes needed to curb disease spread.

NOTES

CHAPTER 1

1. These four lectures have deep roots. The first lecture I gave on the topic of spatial diffusion was part of a course on locational analysis for second-year undergraduates at Cambridge University over the period 1957 to 1966. On moving to Bristol, the course was remodelled and from about 1975 through to my retirement in 1998 it was given as a third-year undergraduate class specifically entitled 'Spatial Diffusion'. As I point out in the Preface, the work reported here represents several decades of work, almost all of it jointly written with a team of whom Andrew Cliff, now professor of theoretical geography at Cambridge, was the leading member. The three most relevant of several jointly written books are: A. D. Cliff, P. Haggett, J. K. Ord, and G. R. Versey, *Spatial Diffusion: An Historical Geography of Epidemics in an Island Community*. Cambridge: Cambridge University Press, 1981; A. D. Cliff, P. Haggett, and M. Smallman-Raynor, *Measles: An Historical Geography of a Major Human Viral Disease. From Global Expansion to Local Retreat, 1840–1990*. Oxford: Blackwell Reference, 1993; A. D. Cliff, P. Haggett, and M. Smallman-Raynor, *Island Epidemics*. Oxford: Oxford University Press, 2000.

2. These ideas are explored at length in Peter Haggett, *The Geographer's Art*. Oxford: Blackwell, 1990.

3. August Lösch, *The Economics of Location*. Yale: Yale University Press, 1954. This was translated from the second revised edition of the German edition by W. H. Woglom. The original German edition was completed in autumn 1939 and published the following year. An appreciation of Losch's geographical work is given by Peter Gould, see 'August Lösch as a child of his times'. In *Becoming a Geographer*. Syracuse, UT: Syracuse University Press, 1999, pp. 273–82.

4. Torsten Hägerstrand, *Innovation Diffusion as a Spatial Process*. Chicago: University of Chicago Press, 1969. This was a translation by Alan Pred of Hägerstrand's original work in Swedish, *Innovationsförloppet ur Korologisk Synpunkt*. Lund: Gleerup, 1953. Hägerstrand's ideas on Monte Carlo simulation were set out in 'On Monte Carlo simulation of diffusion'. In W. L. Garrison and W. F. Marble (ed.), *Quantitative Geography: Part I. Economic and Cultural Topics*. Evanston, Ill.: Northwestern Studies in Geography, 13 (1967), 1–32. A useful summary of his ideas is given in M. H. Yeates, *An*

Introduction to Quantitative Analysis in Human Geography. New York: McGraw Hill, 1974.

5. Gunnar Tornqvist, *Growth of TV Ownership in Sweden, 1956–65.* Uppsala: Uppsala University Press, 1967.

6. E. Casetti and R. K. Semple, 'Concerning the testing of spatial diffusion hypotheses'. *Geographical Analysis*, **1** (1969), 154–9.

7. A. D. Cliff and J. K. Ord, 'Space–time modelling with an application to regional forecasting'. *Institute of British Geographers, Publications*, **64** (1975), 119–28.

8. D. Mollison, 'Spatial contact models for ecological and epidemic spread'. *Journal of the Royal Statistical Society, Series B*, **39** (1977), 283–326.

9. A description of the model is given in P. Haggett, A. D. Cliff, and A. E. Frey, *Locational Analysis in Human Geography* (2nd edn). London: Arnold, 1977, Ch. 7; and in Cliff *et al.* 1981, *op. cit.* [note 1], Ch. 2.

10. Hägerstrand, 1967, *op. cit.* [note 4], 1–32.

11. R. S. Yuill, 'A simulation study of barrier effects in spatial diffusion studies'. *Michigan Inter-University Community of Mathematical Geographers, Discussion Papers*, 5 (1965).

12. R. L. Morrill, 'Waves of spatial diffusion'. *Journal of Regional Science*, **8** (1968), 1–18; id. 'The shape of diffusion in time and space'. *Economic Geography*, **46** (1970), 259–68.

13. The role of urban hierarchies in diffusion processes is explored in Cliff *et al.* 1981, *op. cit.* [note 1], 26–32.

14. This was subsequently published as *Locational Analysis in Human Geography*. London: Arnold, 1965.

15. The work on south-west England was published as P. Haggett, 'Leads and lags in interregional systems: a study of cyclic fluctuations in the South West economy'. In M. D. I. Chisholm and G. Manners (ed.), *Spatial Policy Problems in the British Economy*. Cambridge: Cambridge University Press, 1972, 69–95. A good example of the methodology for studying waves in urban system is L. J. King, E. Casetti, and D. Jeffrey, 'Economic impulses in a regional system of cities'. *Regional Studies*, **3** (1969), 213–18.

16. Brian Berry is the leading urban geographer in the US. He was at the time professor of urban geography in the University of Chicago. After appointment at Harvard and Carnegie–Mellon he is now at the University of Texas.

17. Serendipity is the faculty of making happy and unexpected discoveries by accident. It was coined by Horace Walpole in 1754 in the fairytale, *The Three Princes of Serendip* (Sri Lanka). Its general role in geographical writing has been explored by Gladys Wrigley of the American Geographical Society in 'Adventures in serendipity; thirty years of the "Geographical Review"'. *Geographical Review*, **42** (1952), 511–42.

18. J. Cassel, P. Haggett, H. O. Lancaster, J. Pfanz, C. Vukanovic, and G. S.

Watson, *Advisers' Report on WHO Research in Epidemiology and Communications Science*. Geneva: United Nations, WHO, Paper RECS/69.6, 1969.

19. A. S. Benenson (ed.), *Control of Communicable Diseases in Man* (15th edn). Washington, DC: American Public Health Association, 1990.

20. A. D. Cliff and P. Haggett, *Atlas of Disease Distributions: Analytical Approaches to Epidemiological Data*. Oxford: Blackwell Reference, 1988.

21. R. Stone, 'Guam: deadly disease dying out'. *Science*, **261** (1993), 424–6. For an extensive review see Cliff *et al.* 2000, *op. cit.* [note 1].

22. *The World Health Report 1995: Bridging the Gaps*. Geneva: WHO, 1995.

23. J. V. Bennett, S. D. Holmberg, M. F. Rogers, and S. L. Solomon, 'Infectious and parasitic diseases'. In R. W. Amler and H. B. Dull (ed.), *Closing the Gap. The Burden of Unnecessary Diseases*. Oxford: Oxford University Press, 1987, 100–20.

24. F. L. Black, 'Measles'. In A. S. Evans (ed.), *Viral Infections of Humans*. New York: Plenum, 1984, 397–418; citation on 397.

25. Cliff *et al.* 1993, *op. cit.* [note 1], 6.

26. Cliff *et al.* 1993, *op. cit.* [note 1], 6–7.

27. The most accessible source is Charles Creighton's translation of the second edition as A. Hirsch, *Handbook of Geographical and Historical Pathology*, Vols 1–3. London: New Sydenham Society, 1883.

28. D. Bernoulli, 'Essai d'une nouvelle analyse de la mortalité causée par la vérole et des avantages de l'incubation pour la prévenir' [A new analysis of smallpox mortality and the benefits of incubation for its prevention]. *Mémoires Mathématique et Physique, Académie Royale des Sciences* (Paris), **1** (1760), 1–10.

29. W. Farr, 'Progress of epidemics', *Registrar General of England and Wales, Second Report*, 1840, 91.

30. J. Brownlee, 'Statistical studies in immunity: the theory of an epidemic'. *Proceedings of the Royal Society of Edinburgh*, **26** (1907), 84–521.

31. The range of models used is reviewed in N. T. J. Bailey, *The Mathematical Theory of Infectious Diseases and its Applications*. London: Griffin, 1975, and in R. M. Anderson and R. M. May, *Infectious Diseases of Humans: Dynamics and Control*. Oxford: Oxford University Press, 1991. The classic non-mathematical account of epidemic spread is given by Sir Macfarlane Burnet and David White, *Natural History of Infectious Disease* (4th edn). Cambridge: Cambridge University Press, 1972, 105–36.

32. Standard accounts of the Hamer–Soper models are given in Bailey, 1975, *op. cit.* [note 31]. An elementary introduction is given in P. Haggett, 'Simple epidemics in human populations: some geographical aspects of the Hamer–Soper diffusion models'. In R. F. Peel, M. D. I. Chisholm, and P. Haggett (ed.), *Processes in Physical and Human Geography: Bristol Essays*. London: Heinemann, 1975, 373–91; and in P. Haggett, 'Prediction and predictability in

geographical systems'. *Institute of British Geographers, Publications*, **19** (1994), 161–75.

33. M. S. Bartlett, 'Measles periodicity and community size'. *Journal of the Royal Statistical Society, Series A*, **120** (1957), 48–70; id. 'The critical community size for measles in the United States'. *Journal of the Royal Statistical Society, Series A*, **123** (1960), 37–44.

34. Bartlett, 1960, *op. cit.* [note 33], 37.

35. For a full statement see A. D. Cliff and P. Haggett, 'Spatial aspects of epidemic control'. *Progress in Human Geography*, **13** (1989), 313–47.

36. D. G. Kendall, 'La propagation d'une épidémie au d'un bruit dans une population limité' [The propagation of an epidemic started by random events within a limited population]. *Publications de l'Institute de Statistique de l'Université de Paris*, **6** (1957), 307–11.

37. A. W. Gilg, 'A study in agricultural disease diffusion: the case of the 1970–71 fowl-pest epidemic'. *Institute of British Geographers, Publications*, **59** (1973), 77–97.

38. Hägerstrand, 1967, *op. cit.* [note 4].

39. This section is taken from a fuller discussion in Haggett, 1995, *op. cit.* [note 2], 102–5.

CHAPTER 2

1. This chapter reflects some thirty years of work on epidemics on islands around the world recently summarized in A. D. Cliff, P. Haggett and M. Smallman-Raynor, *Island Epidemics*. Oxford: Oxford University Press, 2000. The field work with Andrew Cliff on Iceland was originally described in A. D. Cliff, P. Haggett, J. K. Ord, and G. R. Versey, *Spatial Diffusion: An Historical Geography of Epidemics in an Island Community*. Cambridge: Cambridge University Press, 1981. That on Fiji in A. D. Cliff and P. Haggett, *The Spread of Measles in Fiji and the Pacific: Spatial Components in the Transmission of Epidemic Waves Through Island Communities*. Department of Human Geography publication HG/18. Canberra: Research School of Pacific Studies, Australian National University, 1985. An abbreviated account of the epidemics in both islands is provided in A. D. Cliff and P. Haggett, *Atlas of Disease Distributions: Analytic Approaches to Epidemiological Data*. Oxford: Blackwell Reference, 1988, 245–58; A. D. Cliff, P. Haggett, and M. Smallman-Raynor, *Measles: An Historical Geography of a Major Human Viral Disease. From Global Expansion to Local Retreat, 1840–1990*. Oxford: Blackwell Reference, 1993.

2. The original report by Peter Panum appeared in 1847 in Danish, 'Iagttagelser anstillede under maeslinge-epidemien paa Faeroerne i Aaret 1846' ['Observations made during the epidemic of measles on the Faeroe Islands in

the year 1846']. *Bibliothek for Laeger*, **1** (1847), 270–344. It has been reprinted in translation several times, e.g. 'Observations made during the epidemic of measles in the Faeroe Islands in the year 1846'. *Medical Classics*, **3** (1939), 803–86.

3. Cliff, *et al.* 2000, *op. cit.* [note 1], Ch. 1.
4. The significance of Darwin's work in islands for biogeography is summarized and extended in R. H. MacArthur, *Geographical Ecology*, New York: Harper & Row, 1972. See also R. H. MacArthur and E. O. Wilson, *The Theory of Island Biogeography*, Mongraphs in Population Biology. Princeton: Princeton University Press, 1967.
5. The range of epidemiological literature is spanned by Panum, 1847, *op. cit.* [note 2] through to such recent molecular studies as D. Guris, S. B. Auerbach, C. Vitek, and E. Maes, 'Measles outbreaks in Micronesia, 1991–1994'. *Pediatric Infectious Diseases Journal*, **17** (1998), 33–9.
6. F. L. Black, 'Measles endemicity in insular populations: critical community size and its implications'. *Journal of Theoretical Biology*, **11** (1966), 207–11.
7. Cliff, *et al.* 1981, *op. cit.* [note 1], 60.
8. Described in Cliff *et al.* 1981, *op. cit.* [note 1], 62–4.
9. The work on the 1904 epidemic was conducted with the help of Rosemary Graham who translated the difficult manuscript records. See A. D. Cliff, P. Haggett, and R. Graham, 'Reconstruction of diffusion processes at different geographical scales: the 1904 measles epidemic in northwest Iceland'. *Journal of Historical Geography*, **9** (1983*a*), 29–46; id. 'Reconstruction of diffusion processes at local scales: the 1846, 1882 and 1904 measles epidemics in northwest Iceland'. *Journal of Historical Geography*, **9** (1983*b*), 347–68.
10. Cliff *et al.* 1993, *op.cit.* [note 1], 90.
11. Cliff *et al.* 1981, *op.cit.* [note 1], 71–6.
12. Cliff *et al.* 1981, *op.cit.* [note 1], 76.
13. A. D. Cliff, P. Haggett, and J. K. Ord, 'Graph theory and geography'. In R. J. Wilson and L.W.Beineke (ed.) *Applications of Graph Theory*. New York: Academic Press, 1976.
14. One of the earliest accounts is given in P. Haggett and R. J. Chorley, *Network Analysis in Geography*. London: Arnold, 1969.
15. The work on modelling measles epidemics in Iceland is the work of Andrew Cliff together with the econometrician Keith Ord. See e.g. A. D. Cliff and J. K. Ord, 'Forecasting the progress of an epidemic'. In R. L. Martin, N. J. Thrift, and R. J. Bennett (ed.), *Towards the Dynamic Analysis of Spatial Systems*. London: Pion, 1978, 191–204; id. A. D. Cliff and J. K. Ord, (1984). 'Forecasting the spread of an epidemic'. In O. D. Anderson, J. K. Ord, and E. A. Robinson (ed.), *Time Series Analysis: Theory and Practice*, Vol. 6. Amsterdam: North-Holland, 1984, 297–308. See also A. D. Cliff, P. Haggett, and J. K.Ord, 'Forecasting epidemic pathways for measles in Iceland: the use

of simultaneous equations and logit models'. *Ecology of Disease*, **2** (1983), 377–96. A summary is provided in Cliff *et. al.* 1993, *op. cit.* [note 1], 393–410.

16. The second half of the lecture is based largely on work first reported in Cliff and Haggett, 1985, *op. cit.* [note 1]. For an excellent introduction to the geography of Fiji see R. G. Ward, *Land Use and Population in Fiji: a Geographical Study*. Department of Technical Cooperation, Overseas Research Publications, 9. London: HMSO, 1965.

17. This is summarized in Cliff and Haggett, 1985, *op. cit.* [note 1], *passim*.

18. Cliff *et al.* 1993, *op. cit.* [note 1], 131.

19. R. B. Joyce, *Sir William McGregor*. Melbourne: Oxford University Press, 1971, 25.

20. Much of the work on Fijian records is due to the late Revd Alan Tippett who was a Wesleyan missionary in Fiji for several decades. See A. R. Tippett, 'Shifting foci of Methodist witness in Fiji, 1835–1900'. In A. R. Tippett (ed.), *Historical Writings, 1947–1967*. Canberra: St Mark's Library, 1953, 1–58; id. *The Ethnology of Depopulation in the Pacific with Special Reference to Fiji*. Canberra: St Mark's Library, 1974.

21. B. G. Corney, 'The behaviour of certain epidemic diseases in natives of Polynesia with special reference to Fiji'. *Epidemiological Society of London Transactions* (*New Series*), **3** (1883–4), 76–94; see 85. See also W. Squire, 'On measles in Fiji'. *Transactions of the Epidemiological Society of London* (Sessions 1875–76 to 1880–81), **4** (1892), 72–4.

22. Corney, 1883–4, *op. cit.* [note 21], 80.

23. Corney, 1883–4, *op. cit.* [note 21], 84.

24. Norma McArthur, *Island Populations of the Pacific*. Canberra: Australian National University Press, 1967.

25. Cliff *et al.* 1993, *op. cit.* [note 1], 136.

26. For a full account with sources see Cliff *et al.* 1993, *op. cit.* [note 1], Table 2.7; 36.

27. Cliff *et al.* 1993, *op. cit.* [note 1], 136.

28. Cliff *et al.* 1993, *op. cit.* [note 1], 137.

29. K. L. Gillion, *Fiji's Indian Migrants: A History to the End of Indenture in 1920*. Melbourne: Oxford University Press, 1962.

30. Cliff *et al.* 1993, *op. cit.* [note 1], 138–40.

31. Cliff *et al.* 1993, *op. cit.* [note 1], 140.

CHAPTER 3

1. This third lecture draws largely on two groups of papers. For the historical work on past origins and dispersals see P. Haggett, 'Sauer's "Origins and Dispersals": its implications for the geography of disease'. *Transactions of the Institute of British Geographers* (*New Series*), **17** (1992), 387–98; id.

'Geographical aspects of the emergence of infectious diseases'. *Geografiska Annaler, Series B*, **76** (1994), 91–104. For current trends and their implications see A. D. Cliff and P. Haggett, 'Disease implications of global change'. In R. J. Johnston, P. Taylor, and M. Watts (ed.), *Geographies of Global Change: Remapping the World in the Late Twentieth Century*. London: Routledge, 1995, 206–23; id. 'Global trends in communicable disease control'. In Norman Noah and Mary O'Mahony (ed.) *Communicable Disease: Epidemiology and Control*. Chichester: John Wiley, 1998, 3–46.

2. L. Garrett, 'The next epidemic'. *AIDS in the World*. Cambridge, MA: Harvard University Press, 1993

3. J. Cairns, *Cancer, Science and Society*. San Francisco: W. H. Freeman, 1975.

4. H. O. Lancaster, *Expectations of Life: A Study of the Demography, Statistics and History of World Mortality*. Berlin: Springer, 1990. Maps of the known hearths of cholera in the nineteenth century are provided in A. D. Cliff and P. Haggett, *Atlas of Disease Distributions: Analytic Approaches to Epidemiological Data*. Oxford: Blackwell Reference, 1988, 3–11.

5. A. D. Cliff, P. Haggett, and J. K. Ord, *Spatial Aspects of Influenza Epidemics*. London: Pion, 1986.

6. The geographical pattern of AIDS in the opening decade of its spread has been tracked in M. Smallman-Raynor, A. D. Cliff, and P. Haggett, *International Atlas of AIDS*. Oxford: Basil Blackwell, 1992.

7. D. Huminer, J. B. Rosenfeld, and S. D. Pitlik, 'AIDS in the pre-AIDS era'. *Review of Infectious Diseases*, **9** (1987), 1102–8.

8. A. S. Benenson, *Control of Communicable Diseases in Man* (15th edn). Washington, DC: American Public Health Association, 1990.

9. N. M. Ampel, 'Plagues—what's past is present: thoughts on the origin and history of new infectious diseases'. *Review of Infectious Diseases*, **13** (1991), 658–65.

10. R. Rosqvist, M. Skurnik, and H. Wolf-Watz, 'Increased virulence of Yersinia pseudotuberculosis by two independent mutations'. *Nature* **334** (1988), 522–4.

11. A. G. Carmichael and A. M. Silverstein, 'Smallpox in Europe before the seventeenth century: virulent killer or benign disease?' *Journal of the History of Medicine and Allied Sciences*, **452** (1987), 147–68.

12. T. F. Smith, A. Srinivasan, G. Schochetman, M. Marcus, and G. Myers, 'The phyogenetic history of immunodeficiency viruses'. *Nature*, **333** (1988), 573–5.

13. A. D. Cliff and P. Haggett, *The Spread of Measles in Fiji and the Pacific: Spatial Components in the Transmission of Epidemic Waves through Island Communities*. Canberra: Australian National University, 1985.

14. T. McKeown, *The Origins of Human Disease*. Oxford: Blackwell, 1988, 4–5.

15. Data is provided in C. Creighton, *A History of Epidemics in Britain* (2 vols). Cambridge: Cambridge University Press, 1894; and in A. B. Janetta, *Epidemics and Mortality in Early-Modern Japan*. Princeton: Princeton

University Press, 1987. The data have been plotted in A. D. Cliff, P. Haggett, and M. Smallman-Raynor, *Measles: An Historical Geography of a Major Human Viral Disease. From Global Expansion to Local Retreat, 1840–1990.* Oxford: Blackwell Reference, 1993, Figs 3.2, 3.4; 53, 57.

16. S. Zivanovic, *Ancient Diseases: The Elements of Palaeopathology.* London: Methuen, 1982, 217–45.

17. A. J. Levine, *Viruses.* New York: Scientific American, 1992, 64.

18. A. D. Langmuir, T. D. Worthen, J. Solomon, C. G. Ray, and E. Petersen, 'The Thucydides syndrome: a new hypothesis for the cause of the plague of Athens'. *New England Journal of Medicine*, 333 (1985), 1027–30; A. J. Holladay, 'The Thucydides syndrome: another view'. *New England Journal of Medicine*, 335 (1986), 1170–3.

19. The paper in question was J. D. Watson and F. H. Crick, 'Molecular structure of nucleic acids: a structure for deoxyribose nucleic acid'. *Nature*, 4356 (1953), 737–8. The story of the research and the race to unravel the structure of DNA is engagingly told in J. D. Watson, in *The Double Helix: A Personal Account of the Discovery of the Structure of DNA.* (A new critical edition edited by G. S. Stent.) London: Weidenfeld & Nicolson, 1981.

20. A useful survey of modern genetics is given by Steve Jones, Reader in Genetics at University College London in his 1992 Reith Lecture on 'The language of the genes'.

21. Jones, 1992, *op. cit.* [note 20], Lecture 3.

22. C. O. Sauer, *Agricultural Origins and Dispersals.* New York: American Geographical Society, 1952.

23. Sauer, 1952, *op. cit.* [note 22], v.

24. 'God helping, I shall never write anything more on methodology: it's a habit forming drug'. C. O. Sauer: Letter to Gladys Wrigley, dated 22 March 1932. Gladys Wrigley was then the influential editor of the *Geographical Review*, a post she held for thirty years. She retired from the American Geographical Society in 1949 and from her home in New Milford, Connecticut, continued to edit some of the Society's occasional publications including Sauer's Bowman Lectures. Quotations from this letter and other correspondence are taken from the Sauer Papers [SP] located in the Bancroft Library, University of California at Berkeley. A useful guide to the papers is provided by M. S. Kenzer, 'Carl Sauer and the Carl Ortwin Sauer Papers'. *History of Geography Newsletter*, 5 (1986), 1–9.

25. C. O. Sauer. Letter to Gladys Wrigley, dated 21 May 1952 [SP].

26. C. O. Sauer, 1952, *op. cit.* [note 22], 106.

27. F. Galton, 'The first steps towards the domestication of animals'. *Transactions of the Ethnological Society of London* (*New Series*), 3 (1865), 122–38.

28. M. Greenwood, *Epidemics and Crowd Diseases.* London: Williams & Norgate, 1935.

29. M. S. Bartlett, 'Measles periodicity and community size'. *Journal of the Royal Statistical Society, Series A*, **120** (1957), 48–70.
30. F. Fenner, 'The effects of changing social organization on the infectious diseases of man'. In S. W. Boyden (ed.), *The Impact of Civilization on the Biology of Man*. Toronto: University of Toronto Press, 1970; R. N. Fiennes, *Zoonoses and the Origins and Ecology of Infectious Disease*. London: Academic Press, 1978.
31. N. D. Cook and W. G. Lovell (ed.), *Secret Judgments of God: Old World Disease in Colonial Spanish America*. Norman, OK: University of Oklahoma Press, 1992, 287–94.
32. A. D. Cliff, P. Haggett, and M. Smallman-Raynor, *Island Epidemics*. Oxford: Oxford University Press, 2000.
33. G. F. Pyle, 'Diffusion of cholera in the United States'. *Geographical Analysis*, **1** (1969), 59–75.
34. Cliff & Haggett, 1985, *op. cit.* [note 13], 60–4.
35. Cliff *et al.* 1993, *op. cit.* [note 15], 46–52.
36. T. Sadik, *The State of World Population 1991*. New York: United Nations Population Fund, 1991.
37. P. Haggett, 'Some components of global environmental change'. In B. A. Bannister (ed.), *Report of a Think Tank on the Potential Effects of Global Warming and Population Increase on the Epidemiology of Infectious Diseases*. London: Public Health Laboratory Service, 1991, 5–14.
38. B. Williams, 'Assessing the health impact of urbanization'. *World Health Statistics Quarterly*, **43** (1989), 145–52.
39. D. W. T. Crompton and L. Savioli, 'Terrestrial parasitic infections and urbanization'. *Bulletin of the World Health Organization*, **71** (1993), 1–7.
40. P. Haggett, 'Sauer's "Origins and dispersals": its implications for the geography of disease'. *Transactions, Institute of British Geographers*, **17** (1992), 387–98.
41. P. E. M. Fine, 'Herd immunity. History, theory, practice'. *Epidemiologic Reviews*, **15** (1993), 265–302.
42. Peter Gould, *Becoming a Geographer*. Syracuse, UT: Syracuse University Press, 1999.
43. G. P. Schmid, 'The global distribution of Lyme disease'. *Reviews of Infectious Diseases*, **7** (1985), 41–50.
44. H. B. Smith and D. Tirpak (ed.), *Potential Effects of Global Climatic Change on the United States: Vol. G. Health*. Washington, DC: US Government Printing Office [Environmental Protection Agency, EPA 230-05-89-057], 1989. C. Ewan, E. Bryant, and D. Calvert, *Health Implications of Long Term Climate Change*. Canberra: National Health and Medical Research Council of Australia, 1990. B. A. Bannister (ed.), *Report of a Think Tank on the Potential Effects of Global Warming and Population Increase on the Epidemiology of Infectious Diseases*. London: Public Health Laboratory Service, 1991.

45. Intergovernmental Panel on Climatic Change (IPCC). *Scientific Assessment of Climate Change*. Report prepared for IPCC by Working Group 1. World Meteorological Organization, Geneva, 1990.
46. A. Henderson-Sellars and R. J. Blong, *The Greenhouse Effect: Living in a Warmer Australia*. Kensington, Sydney: New South Wales University Press, 1989.
47. D. J. Bradley, 'The scope of travel medicine'. In R. Steffen *et al.* (ed.), *Travel Medicine: Proceedings of the First Conference on International Travel Medicine, Zurich, Switzerland, April 1988*. Berlin: Springer, 1988, 1–9.
48. R. Steffen and H. O. Lobel. In *International Travel and Health: Vaccination Requirements and Health Advice*. Geneva: WHO, 1995, 56.
49. M. Bouvier, D. Pittet, L. Loutan, and M. Starobinski, 'Airport malarias: a mini-epidemic in Switzerland'. *Schweizerin Medizin Wochenschreiber*, **120** (1990), 1217–22.
50. Bradley, 1988, *op. cit.* [note 47], 4.
51. M. Khlat and M. Khoury, 'Inbreeding and diseases: demographic, genetic and inbreeding perspectives'. *Epidemiological Reviews*, **13** (1991), 28–41.
52. Smallman-Raynor *et al.* 1992, *op. cit.* [note 6], 132–5.

CHAPTER 4

1. This last lecture is a development of one originally written as part of the centenary programme to mark the 100th anniversary of the Cambridge Department of Geography. It was given at Cambridge 3 November 1989 and published as A. D. Cliff and P. Haggett, 'Spatial aspects of epidemic control'. *Progress in Human Geography*, **14** (1989), 315–47. Since then it has been expanded and developed in a number of forms; see A. D. Cliff and P. Haggett, 'Global trends in communicable disease control'. In Norman Noah and Mary O'Mahony (ed.), *Communicable Disease: Epidemiology and Control*. Chichester: John Wiley, 1998, 3–46; and with applications specifically to measles in the final chapter of A. D. Cliff, P. Haggett, and M. Smallman-Raynor, *Measles: An Historical Geography of a Major Human Viral Disease. From Global Expansion to Local Retreat, 1840–1990*. Oxford: Blackwell Reference, 1993, 413–26.
2. N. T. J. Bailey, *The Mathematical Theory of Infectious Diseases and Its Applications*. London: Griffin, 1975.
3. E. H. Hinman, *World Eradication of Infectious Diseases*. Springfield, IL: Thomas, 1966. W. W. Spink, *Infectious Diseases: Prevention and Treatment in the Nineteenth and Twentieth Centuries*. Folkestone, UK: Dawson, 1978.
4. Cliff and Haggett, 1989, *op. cit.* [note 1], 316–18.
5. R. R. Tinline, A Simulation Study of the 1967–68 Foot-and-Mouth Epizootic in Great Britain. Unpublished D.Phil. dissertation. Department of Geography, University of Bristol, England, 1972.

6. M. Burnet and D. O. White, *Natural History of Infectious Diseases.* Cambridge: Cambridge University Press, 1972

7. F. L. Black, 'Measles endemicity in insular populations: critical community size and its implications'. *Journal of Theoretical Biology*, **11** (1966), 207–11.

8. D. Greenhalgh, 'Simple models for the control of epidemics', *Mathematical Modelling*, **7** (1986), 753–63.

9. D. A. Griffiths, 'The effects of measles vaccination on the incidence of measles in the community'. *Journal of the Royal Statistical Society, Series A*, **136** (1973), 441–9; R. M. Anderson and B. T. Grenfell, 'Quantitative investigations of different vaccination policies for the control of congenital rubella syndrome (CRS) in the United Kingdom'. *Journal of Hygiene*, **96** (1986), 305–33; R. M. Anderson and R. M. May, 'Vaccination against rubella and measles: quantitative investigation of different policies'. *Journal of Hygiene*, **90** (1983), 295–325.

10. Bailey, 1975, *op. cit.* [note 2], 358–70; B. Cvjetanovic, B. Grab, and H. Dixon, 'Epidemiological models of poliomyelitis and measles and their application to the planning of immunization programmes'. *Bulletin of the World Health Organization*, **60** (1982), 405–22; S. Greenland and R. R. Frerichs, 'On measures and models for the effectiveness of vaccines and vaccination programmes'. *International Journal of Epidemiology*, **17** (1988), 456–63; E. G. Knox and H. S. Shammon, 'A model basis for the control of whooping cough'. *International Journal of Epidemiology*, **15** (1986), 452–60.

11. R. M. Anderson and D. J. Nokes, 'Mathematical models of transmission and control'. In W. W. Holland, R. Detels, and G. Knox (ed.), *Oxford Textbook of Public Health* (2nd edn). Oxford: Oxford University Press, 1991, 225–51.

12. The evolution of Centers for Disease Control (CDC) from a Second World War agency established to fight malaria in the American South is traced by Elizabeth W. Everidge, *Sentinels for Health: A History of the Centers for Disease Control*. Berkeley: University of California Press, 1992.

13. R. W. Sutter, S. E. Markowitz, J. M. Bennetch, W. Morris, E. R. Zell, and S. R. Preblud, 'Measles among the Amish: a comparative study in primary and secondary cases in households'. *Journal of Infectious Diseases*, **163** (1991), 12–16. See also CDC, 'Outbreaks of rubella among the Amish—United States, 1991'. *Morbidity and Mortality Weekly Report*, **40** (1991), 264–5. The background to Amish settlement in the US is given in W. K. Crowley, 'Old order Amish settlement: diffusion and growth'. *Annals of the Association of American Geographers*, **68** (1978), 249–64.

14. K. F. Kiple (ed.), *The Cambridge World History of Human Disease.* Cambridge: Cambridge University Press, 1993, 198.

15. M. I. Roemer, 'Internationalism in medicine and public health'. In W. F. Bynum and R. Porter (ed.), *Encyclopaedia of the History of Medicine*, Vol. 2. London: Routledge, 1993, 1417–35.

16. The impact of isolation on the incidence of epidemic diseases in Australia has been examined by H. O. Lancaster, 'Deafness as an epidemic disease in Australia. A note on census and institutional data'. *British Medical Journal*, **2** (1951), 1429–32; id. 'The epidemiology of deafness due to maternal rubella'. *Acta Genetica*, **5** (1954), 12–24; id. 'The infections and population size in Australia'. *Bulletin of the International Statistical Institute*, **42** (1967), 459–71. See also the reviews of Australian quarantine policy in J. H. L. Cumpston, *Influenza and Maritime Quarantine in Australia.* Sydney: Commonwealth of Australia, Quarantine Service Publication 18, 1919; id. *The History of Diphtheria, Scarlet Fever, Measles, and Whooping Cough in Australia, 1788–1925.* Melbourne: Commonwealth of Australia, Department of Health Service Publication 37, 1927.

17. F. G. Ball, 'Spatial models for the spread and control of rabies incorporating group size'. In P. J .Bacon (ed.) *Population Dynamics of Rabies in Wildlife.* London: Academic Press, 1985, 197–222.

18. F. Steck and A. Wandeler, 'Epidemiology of fox rabies in Europe'. *Epidemiological Reviews*, **2** (1980), 71–6.

19. D. W. MacDonald, *Rabies and Wildlife: a Biologist's Perspective.* Oxford: Oxford University Press, 1980.

20. K. Kuulusmaa, 'The spatial general epidemic and locally dependent random graphs'. *International Journal of Applied Probability*, **19** (1982), 745–58; D. Mollison, 'Spatial contact models for ecological and epidemic spread'. *Journal of the Royal Statistical Society, Series B*, **39** (1977), 283–326.

21. A. Kallen, P. Arcuri, and J. D. Murray, 'A simple model for the spatial spread and control of rabies'. *Journal of Theoretical Biology*, **116** (1985), 377–93.

22. MacDonald, 1980, *op. cit.* [note 19], 110–11.

23. P. R. Gould, 'La géographie du Sida: étude des flux humains et expansion d'une épidémie' [The geography of AIDS: study of human movement and the expansion of an epidemic]. *La Recherche*, **280** (1995),36–7.

24. The Russian models produced by Baroyan and his colleagues are reviewed in A. D. Cliff, P. Haggett, and J. K. Ord, *Spatial Aspects of Influenza Epidemics.* London: Pion, 1986, 39–44.

25. A. D. Cliff and P. Haggett, *Atlas of Disease Distributions: Analytical Approaches to Epidemiological Data.* Oxford: Blackwell Reference, 1988.

26. Tinline, 1972, *op. cit.* [note 5], 274–325.

27. F. Fenner, D. A. Henderson, I. Arita, Z. Jesek, and I. D. Ladnyi, *Smallpox and its Eradication.* Geneva: WHO, 1988.

28. A. D. Cliff and P. Haggett, *Atlas of Disease Distributions: Analytical Approaches to Epidemiological Data.* Oxford: Blackwell, 1988, 223–8.

29. F. G. Cutts and P. G. Smith, *Vaccination and World Health.* London: Wiley, 1994; Global Programme for Vaccines and Immunization, *Immunization Policy.* Geneva: WHO, 1995.

30. F. Fenner 'The eradication of infectious diseases'. *South African Medical Journal*, **66** suppl. (1986), 35–9.

31. The map was redrawn from data provided in a WHO publication: 'La poliomélite sera éradiquée' [Poliomyelitis will be eradicated]. Geneva; WHO, 7 April 1995, in Cliff and Haggett, 1998, *op. cit.* [note 1], Fig. 1.13, 41.

32. G. W. Matthews and R. E. Churchill. 'Public health surveillance and the law'. In S. M. Teutsch and R. E. Churchill (ed.), *Principles and Practice of Public Health Surveillance*. New York: Oxford University Press, 1994, 190–9.

34. D. F. Stroup, M. Wharton, K. Kafadar, and A. G. Dean, 'An evaluation of a method for detecting aberrations in public health surveillance data'. *American Journal of Epidemiology*, **137** (1993), 45–9.

34. *The World Health Report 1995: Bridging the Gaps*. Geneva: WHO, 1995, 1.

INDEX

acute lower respiratory infections (ALRI) 13
adoption of pasture subsidy 9
Africa 68, 84,
AIDS, epidemic, United States 88–9, 116; origins 72
Alaska 60
Americas 60
Amish communities 109
Amler, R.W. 133
Ampel, N.M. 73, 137
Anderson, O. D. 135
Anderson, R. M. 21, 104, 106, 133, 141
Anderson–May model 105
Argentine HF 84
Arita, I. 142
Asia 68
Auerbach, S. B. 135
autoregressive models 49

Bailey, N. T. J. 10, 21, 99, 133, 140–1
Ball, F. G. 142
Bannister, B. A. 139
Bartlett: and city size effects 23; model of spatial spread 25
Bartlett, M. S. 22, 32, 53, 82, 102, 134, 139
Bates, Marston 9
Bayesian entropy model 49
Benensen, A. S. 11, 133, 137
Bennetch, J. M. 141
Bennett, J. V. 15, 133
Bennett, R. J. 135
Bernoulli, D. 21, 133
Berry, B. J. L. 9–10, 132
Black Death 11
Black model of epidemic frequency 33
Black, F. L. 33–4, 103, 133, 135, 141
Blong, B. 40
Bolivian HF 84
Bouvier, M. 140
Bradley, D. J. 93–4, 96, 140
Britain: rabies spread, 114
Brownlee, J. 21, 133
Bryant, E. 139
Burnet, Sir Macfarlane 19, 102–3, 133, 141
business depressions 4
Bynum, W. F. 141

Cairns, J. 67, 137
Cakobau, Fiji 58
Calvert, D. 139
Caribbean 68, 84
Carmichael, A.G. 73, 137
Casetti, E. 5, 132
Cassel, J. 132
CDC 141
cerebrovascular disease 13
chain binomial model 47
Chapin 17
China 71, 84
Chisholm M. D. I., 132–3
chlamydial infections 13
cholera, 3; pandemics 70–1
Chorley, R. J. 135
chronic obstructive pulmonar disease 13
Churchill, R. E. 143
Cliff, A. D. 3–5, 11–12, 19–20, 23, 25, 28, 33, 35–6, 38, 40, 43–4, 47, 49–50, 52, 54–5, 60, 62, 71, 75, 83, 100–1, 108, 112, 115, 131–40, 142
communicable diseases, burden of 12–16
containing epidemics 99–130
containment, offensive 111–15
control strategies 99–103
Cook, N. D. 81, 139
Corney, B. G. 58, 136
Creighton, C. 75, 137
Crick, F. H. 138
Crompton, D. W. T. 86, 139
Crowley, W. K. 141
Cumpston, J. H. L. 142
Cutts, F. G. 105, 121, 142
Cvjetanovic B, 141

Darwin, Charles 31, 135
Dean, A. G. 143
dengue 68, 84
Denmark, measles cycles 20
Detels, R. 141
diarrhoea 13
diffusion waves 1–30 passim
diphtheria 123
disease: origins 69–73; dispersals 68–98 passim; burden of 12–16
Dixon, H. 41

Dull, H. B. 133
dysentry 11

East Africa: Marburg fever 97
Ebola fever 68
El Tor 3
elimination campaigns 121–6
encephalitis 84
Enders, J. F. 16, 127
En'ko, P. D. 19, 21
England 11, 38
epidemics: containment of 99–130 *passim*; as
    diffusion waves 1–30 *passim*; disease
    models, historical note 21; historical
    examples 11; nature of 10–12
Europe 11, 68
Everidge, E. W. 141
evidence, limits of 74–77
Ewan, C. 139

Faeroes 38, 60
famine 11
Farr, W. 21, 133
Fenner, F. 81, 119, 122, 124, 139, 124–5,
    142–3
Fiennes, R. N. 81, 139
Fiji 11, 60; 1875 measles outbreak 53–7; 1875
    outbreak, demographic impact of 59–61;
    Indian migration to 61–5; wave sequence
    52
Fine, P. E. M. 139
foot-and-mouth (FMD) epizootic 118–19
France, population mobility 94
Frerichs, R. R. 141
Frey, A. E. 132
Frost 21

Galton, F. 80, 138
Garrett, L. 137
Garrett, L. 137
Garrison, W. L. 131,
genital warts 13
geographical space, collapse of 93–5
geography and spatial diffusion 1–4
Gilg, A.W. 26–28,134
Gillion, K.L. 62, 136
GLIM 49
global change: disease implications 82–5;
    global land use, changes in 87; global
    smallpox camapaign 122; global
    warming 91–3
gonococcal infections 13
Gould, P. R. 87, 89, 113, 115–16, 131, 139,
    142
Grab, B. 141
Graham, R. 135

Greenhalgh, D. 141
Greenland, S. 141
Greenwood, M. 138
Gregg, J. 111
Grenfell, B. T. 141
Griffiths, D. A. 104, 141
Guris, D. 135

Hagerstarnd's Monte Carlo diffusion model
    4–9
Hagerstrand, T. 4, 9, 115, 117, 131–2, 134
Haggett, P. 11, 19, 25, 28, 35–6, 44, 52, 60, 62,
    68, 71, 84–5, 87, 94, 101, 108, 112, 115,
    131–40, 142
Haiti 11
Hamer-Soper models 19, 22, 47, 100, 104, 133
Hantaan virus 68, 84
Hantaviruses 68
Henderson, D. A. 142
Henderson-Sellars, A. 140
hepatitis B 123
Hinman, E. H. 101, 140
Hirsch, A. 21, 133
HIV 68
Holladay, A. J. 138
Holland, W. W. 141
Holmberg, S. D. 133
Hong Kong, influenza 71
HTLV 68
Huminer, D. 137

Iceland 60; disease records 37–43; as graph
    46–7; as laboratory, 33–7; lag structure
    for epidemics 44–5; measles cycles 20;
    medical districts 36; predictive wave
    models 47–52; seasonal distribution of
    measles 50–1; settlement map 35; time
    intervals between epidemics 44; wave
    sequences 43–6;
immunization, impact on epidemic cycles 105
India 84
India-Fiji immigration route 62
infection, natural breaks in 103–4
influenza 11, 68, 84; pandemics 71
Intergovernmental Panel on Climate Change
    (IPCC) 91, 140
Iowa 4
Ireland 11
ischaemic heart disease 13
islands, as laboratories 31–3; epidemics on
    31–65 *passim*
isolation, as defensive strategy 110–11

Janetta, A.B. 75, 137
Japan, influenza 71; measles records 75
Jeffrey, D. 132

Jesek, Z. 142
Johnston, R. J. 137
Jones, S. 77, 138
Jonsson, J. 37
Joyce, R. B. 136
Junin fever 68

Kallen 115
Kalman filter model 49
Kendall model of wave propagation 27
Kendall waves 26–9
Kendall, D. G. 26–9, 134
Kenzer, M. S. 138
Kermack 1
Khlat, M. 96, 140
Khoury, M. 140
King, L. J. 132
Kiple, K. F. 141
Knox, E. G. 141
Knox, G. 141
Kuulusmaa, K. 142
Kyasanur forest disease 84

Ladnyi, I. D. 142
Lancaster, H. O. 10, 69, 111, 132, 137, 142
Langmuir, A. D. 76, 138
Lassa fever 68
Leishmaniasis 84
Levine, A. J. 138
Lobel, H. 95, 140
logistic transformation model 48
London, measles deaths 75
Losch, A. 2, 4, 9, 131
Loutan, L. 140
Lovell, W. G. 81, 139
Lyme disease 84

McArthur, N. 59–60, 136
MacArthur, R. H. 135
MacDonald, D. W. 105, 107, 113, 142
McGregor, W. 56
Machupo fever 68
McKendrick 21
McKeown, T. 74, 137
Maes, E. 135
malaria 13, 84
Manners, G.
map sequences 30
Marble, W. F. 131
Marburg fever 67, 68, 97
Marcus, M. 137
Markowitz, S. E. 141
Martin, R. L. 135
mass-action models 21–6
Matthews, G. W. 143
May, R. M. 21, 133, 141

measles 11, 13, 123, 125; cycles and population size 20; disease 16–21; hearth in Middle East 83; epidemics 19; as tracker Fiji 52–65; United States 108
Middle East: as measles hearth 83
modelling, AIDS epidemic 116
Mollison, D. 6, 113, 132, 142
Monkeypox 84
Monte Carlo methods 6–9
Morris, W. 141
Morse, S. S. 68, 84
mumps epidemics 106
Murray 111
Myers, G. 137

New South Wales: rubella outbreaks 112
New World: agricultural origins 79
Newcastle disease 28
Noah, N. 137, 140
Nokes, D. J. 104, 106, 141
North-West Iceland epidemic map 40
Norway 38

O'Mahony, M. 137, 140
occupational diseases 13
occupational injuries 13
Ord, J. K. 5, 131, 134–5, 142
Oropouche 84

Pacific Basin, measles spread in 54–5
Pacific, measles invasions 52–3
Pacon, P. J. 142
Palestine 11
Panama 84
Panum, P. 31–2, 37, 60, 134–5
Paris 11
Peebles, T. C. 16, 133
Peel, R. F. 133
Pennsylvania, measles outbreak in Amish community 109
Persian Gulf 84
pertussis 123
Petersen, E. 138
Pfanz, J. 132
Pitlik, S. D. 137
Pittet, D. 140
poliomyelitis 122, 125; elimination 126–9
population: global growth 85–7; latitiudinal distribution 85
population mobility, France 94
Porter, R. 141
Preblud, S. R. 141
Pred, A. 131
Pyle, G. F. 81, 139

rabies spread, Britain 114

Ray, C. G. 138
Reed 21
Reykjavik 47–8
Rift Valley fever 68, 84
ring control 115–19
Robbins, W. R. 127
Robinson, E. A. 135
Roemer, M. I. 141
Rogers, M. F. 133
Rosenfeld, J. B. 137
Rosqvist, R. 73, 137
Ross, Sir Ronald 21
rubella, New South Wales 112

Sabin, A. B. 127
Sadik, T. 139
Salk, J. E. 127
Sauer, C.O. 75, 77, 98, 136, 138–9; disease
    origins 80–2; geographical speculation
    77–80
Savioli, L. 86, 139
Scarlet fever 11
Schmid, G. P. 139
Schochetman, G. 137
Semple, R. K. 5, 132
Seoul virus 68
Seoul-like viruses 84
Shammon, H. S. 141
Shannon-Pyle model 72
Silverstein, A. M. 73, 137
simultaneous equation model 49
Singapore, influenza 71
Sink, W. W. 101, 140
Skurnik, M. 137
Smallman-Raynor, M. 11, 72, 97, 131, 134,
    137–40
smallpox eradication, global 119–22, 124–5
Smith 105
Smith, H. B. 139
Smith, P. G. 121. 142
Smith, T. F. 73, 137
Solomon, J. 138
Solomon, S. L. 133
Soper, W. 21
South America 68, 84
space, collapse of 93–5
Squire, W. 136
Srinivasan, A. 137
Starobinski, M. 140
Steck, F. 142
Steffen, R. 95, 140
Stewart, W. H. 67
Stone, R. 133
Stroup, D. F. 143
subacute sclerosing panencephalitis (SSPE)
    108

Sutter, R. W. 109, 141
Sweden 9

Taylor, P. 137
tetanus 123
Teutschm, S. M. 143
Thrift, N. J. 135
Tinline, R. R. 102, 111, 119, 136, 140,
    142
Tippett, A. R. 136
Tirpak, D. 139
Tornqvist, G. 132
travel, increase by generations 93
travellers, disease threats 95
trichomoniasis 13
tropics, disease threats to travellers, 95
tuberculosis 13, 123
typhus 11

Uganda, Marburg fever 97
United Kingdom, measles cycles 20, 38
United States 38, 84; AIDS spread 88–89,
    116; infectious diseases 15; measles
    cycles 20; measles vaccinations 108

vaccination, impacts on epidemic cycles
    104–10
Vavilov, views on plant origins 78
Venezuelan equine fever 84
Versey, G. R. 131, 134
Vitek, C. 135
Vukanovic, C. 132

Wales 11
Walpole, Hugh 132
Wandeler, A. 142
Ward, R. G. 136
Watson, G. S. 10, 133
Watson, J. D. 138
Watts, M. 137
Western Australia 60
Western Europe 11
Wharton, M. 143
White, D. 19, 133, 141
whooping cough 13
Williams, B. 139
Wilson, E. O. 135
Wolf-Watz, H. 137
World Health Organization (WHO) 13, 128;
    WHO, Expanded Programme on
    Immunization 123; Global Programme
    for Vaccines and Immunization 123
Worthen, T.D. 138
Wrigley, G. 80, 132, 138

Yeates, M. H. 9, 131

yellow fever 11, 68, 84, 123
Yuill, R. S. 7, 132

Zell, E. R. 141
Zivanovic, S. 138